D0010601

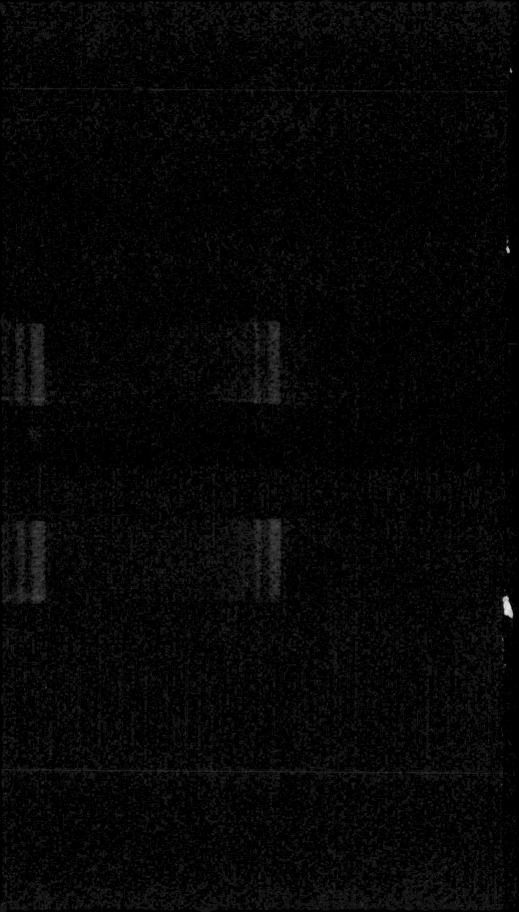

THE J. PAUL GETTY MUSEUM

· HANDBOOK ·

OF THE COLLECTIONS

THE J. PAUL GETTY MUSEUM
· HANDBOOK ·
OF THE COLLECTIONS

THE J. PAUL GETTY MUSEUM
MALIBU · CALIFORNIA
1991

Published 1986. Second edition 1988. Revised edition 1991.
©1986, 1988, 1991 The J. Paul Getty Museum
17985 Pacific Coast Highway
Malibu, California 90265-5799

Mailing address:
P.O. Box 2112
Santa Monica, California 90407-2112

(213) 459-7611 administrative offices
(213) 458-2003 parking reservations

Christopher Hudson, head of publications
Andrea P. A. Belloli, consulting editor
Cynthia Newman Helms, managing editor
Karen Schmidt, production manager
Leslee Holderness, sales and distribution manager

Project staff:
Andrea P. A. Belloli, manuscript editor
Mary Alice Cline, photograph coordinator
Patrick Dooley, designer
Thea Piegdon, production assistant
Karen Schmidt, production manager
Stephenie Blakemore, Donald Hull, Louis Meluso,
    Thomas Paul Moon, Charles Passela, Penelope Potter,
    Ellen Rosenbery, and Jack Ross, photographers

Typography by Andresen Typographics, Los Angeles
Printed by Nissha Printing Co., Ltd., Kyoto, Japan
Library of Congress Cataloging-in-Publication Data

J. Paul Getty Museum.
    The J. Paul Getty Museum handbook of the collections.—Rev. ed.
        p.   cm.
    ISBN 0-89236-189-1 (pbk.)
    ISBN 0-89236-191-3 (hardback)
    1. J. Paul Getty Museum—Catalogs. 2. Art—California—
Malibu—Catalogs. I. Title.
N582.M25A627   1991
708.194'93—dc20   90-5357
                                                          CIP
Cover: Pontormo, Italian, 1494–1557. *Portrait of Cosimo I de' Medici,*
ca. 1537. 89.PA.49.

CONTENTS

FOREWORD   vii

J. PAUL GETTY AND HIS MUSEUM   1

THE ART COLLECTIONS

    ANTIQUITIES   17

    MANUSCRIPTS   65

    PAINTINGS   89

    DRAWINGS   131

    DECORATIVE ARTS   155

    SCULPTURE AND WORKS OF ART   199

    PHOTOGRAPHS   221

FOREWORD

"If a museum is alive and thriving, it is bound to keep changing and improving. This is good for our visitors and good for our staff. But it means that guidebooks go quickly out of date." Thus wrote my predecessor, Stephen Garrett, in a guidebook of 1982 which obeyed his rule and went quickly out of date. So did the *Handbook* we produced in 1986 and reissued in 1988, and already we need an entirely new edition. May it soon be obsolete!

The Getty Museum's collection has been growing steadily, and the galleries are rearranged frequently, so a room-by-room commentary would be short-lived. This *Handbook* is therefore not a guide but instead a selection of important works from the collection with some interpretive remarks by the curators. By way of introduction we provide a concise account of Mr. Getty's life as a collector and of the Museum's brief but remarkable history. The *Handbook* has a companion, the *Guide to the Villa and Its Gardens* (1988), that treats the building—its design, construction, and historical precedents—in some detail.

It has been seventeen years since the present Museum building was opened to the public. The building has weathered gently, the bright colors of the painted decorations in Roman style have softened, and the trees and shrubs have grown up. Our own perceptions have changed as well, since contemporary "postmodern classicism" has made the re-creation of a Roman villa a less outlandish notion than it seemed to many in 1974. Inside the Museum the changes are more striking, for virtually every gallery has been renovated, and most works of art the visitor sees have been bought during the last decade. In that decade the Getty Museum has changed profoundly in other ways as well.

J. Paul Getty's huge legacy gave the trustees a chance to undertake much more than the improvement of the Getty Museum, which in any event could never have absorbed all the income. In 1983 they formed the J. Paul Getty Trust, a private operating foundation (as distinct from a grant-making foundation) devoted to the visual arts. After two years of consultations with professionals in museums, universities, and schools here and abroad, the officers of the Trust developed a program to employ Mr. Getty's endowment for the maximum public benefit. The primary goals of the Trust, in addition to the continued growth of the J. Paul

View of the Main Peristyle Garden and main façade, The J. Paul Getty Museum, 1980. Photo: Julius Shulman.

Getty Museum, are the advancement of the study of the history of art and related disciplines; the improvement of the documentation of art and of tools for scholarly research; the strengthening of the study, practice, and documentation of art conservation; and the fostering of wider and better education in the arts in American schools. The current activities of other organizations of the Trust are described in the essay that follows entitled "J. Paul Getty and His Museum."

The J. Paul Getty Museum has had opportunities for expansion and improvement that could not have been imagined when it opened in 1974. Hundreds of important new works of art have been acquired in the areas of the Museum's three traditional interests: antiquities, French furniture and decorative arts, and European paintings. Classical antiquities, the largest of the Museum's original collections, include important groups of Greek vases, Cycladic figures and vessels, and Hellenistic metalwork. Mr. Getty had a particular passion for the exquisitely made furniture and decorative objects of eighteenth-century France, of which many more have been added in recent years. European paintings in the collection now include several dozen examples of very great importance, among them a group of fine seventeenth-century Dutch paintings. There are also Italian, French, and Spanish works of great distinction and a small but brilliant group of Impressionist and Post-Impressionist pictures. More than a hundred paintings have joined the collection since 1983.

Entire new collections have been formed in the past decade and four new curatorial departments created. The Museum has bought drawings since 1981; there are now nearly three hundred examples of remarkably high quality, and the collection ranks among the finest for its size in the world. In 1983 the Museum entered the field of illuminated manuscripts with the purchase of the most important group of Medieval and Renaissance manuscripts in private hands; since then, some forty acquisitions have made the Getty collection far more diverse and important. In 1984 it was possible to seize another exceptional opportunity and acquire a number of the best private holdings of photographs anywhere. Together with thousands of acquisitions made since then, they now form the finest museum collection of photographs in this country. All of the drawings, manuscripts, and photographs are available for study by students and scholars, and a selection of each is on view in changing exhibitions. The Museum also has been acquiring sculpture and works of art from European countries other than France. Important statues of the sixteenth through the nineteenth centuries can now be seen together with maiolica, glass, goldsmiths' work, and distinguished Italian furniture.

Hardly a room in the Villa or in Mr. Getty's house (now our office annex) has escaped the renovations necessitated by a growing collection and staff. New conservation studios and labs have been built, dozens of offices created, and many entire departments relocated. Storerooms, photo studios, security facilities, and a dozen other areas invisible to the visitor have been renovated. The Museum's gardens have been made even more enchanting by a program of replanting. Security and emergency preparedness have been greatly increased. As the collections and facilities have improved, so have the services available to visitors. We now offer information about the collection through many different means: introductory talks, lectures, publications, labels, didactic shows, and interactive videodisc programs.

Our aims are to satisfy every visitor's curiosity and to make a visit to the Getty Museum as much a memorable adventure for the mind as a delight for the senses. Rather than amass a large, general collection, we will continue to build one of modest scope but, we hope, great distinction. Our traditional emphases on conservation and publications will be maintained. A new Getty Museum building—together with facilities for other activities of the Trust, including the Getty Center for the History of Art and the Humanities and the Getty Conservation Institute—is under construction in Brentwood, about twenty minutes' drive from the present Museum, in the foothills of the Santa Monica Mountains. This building, designed by Richard Meier, will house the collections from the Middle Ages through the nineteenth century. After 1996, when the new Museum is scheduled to open, the building in Malibu will be renovated to serve as America's only museum devoted to Greek and Roman art.

We hope that our visitors not only will enjoy the pleasures of the collection and its setting but also will share some of our exhilaration at the opportunities the Museum has in its future.

John Walsh
Director

Figure 1. Yousuf Karsh, Canadian, b. 1908. J. Paul Getty, *1964.* ©*1964, Yousuf Karsh.*

J. PAUL GETTY AND HIS MUSEUM

J. Paul Getty combined the lives of oil-field wildcatter, shrewd and spectacularly successful businessman, writer, and member of the international art world. His attitude toward art and collecting was complex. Although he maintained that "fine art is the finest investment," he also felt that "few human activities provide an individual with a greater sense of personal gratification than the assembling of a collection of art objects that appeal to him and that he feels have a true and lasting beauty."

When he died on June 6, 1976, at the age of eighty-three, Mr. Getty had collected art for over forty-five years. He was not an obsessive collector like William Randolph Hearst or Joseph Hirshhorn, but collecting was a part of his life that he relished and about which he felt very deeply. In later years he sometimes called himself an addict who could not stop himself from buying works of art. He frequently wrote about the joys of collecting, as in an essay published in 1965:

> I am convinced that the true collector does not acquire objects of art for himself alone. His is no selfish drive or desire to have and hold a painting, a sculpture, or a fine example of antique furniture so that only he may see and enjoy it. Appreciating the beauty of the object, he is willing and even eager to have others share his pleasure. It is, of course, for this reason that so many collectors loan their finest pieces to museums or establish museums of their own where the items they have painstakingly collected may be viewed by the general public.

J. Paul Getty (fig. 1) was born in 1892 in Minneapolis, Minnesota. His father, George F. Getty, was a successful attorney and already wealthy when he entered the oil business in Oklahoma in 1903. The elder Getty moved his family to Southern California in 1906. An only child, young Paul loved to read and enjoyed such sports as swimming and boxing. His father planned to have him enter the family business and encouraged him from the age of sixteen to work summers in the Oklahoma oil fields.

In 1909, after graduating from high school, Paul Getty made his first visit to Europe with his mother and father. He remembered that as part of the tour they visited the Louvre and the National Gallery in London; later he admitted that neither had made much of an impression on him at the time. After a semester at the University of Southern California and three terms at the University of California, Berkeley, he made a trip to China and Japan in May–June 1912, returning with a few pieces of carved ivory, bronzes, and lacquer work—a foretaste of his later passion for collecting. In November 1912 Mr. Getty returned to Europe as a student at

Figure 2. Bernard Molitor, French, 1730–ca. 1815. Roll-Top Desk, ca. 1790. Mahogany veneer; gilt bronze mounts, 136.5 x 177.8 x 86.3 cm (53¼ x 70 x 34 in.). 67.DA.69.

Oxford University to read political science and economics, with a view to entering the diplomatic service. He received his diploma in June 1913 and set off on a *Wanderjahr.* In later years Mr. Getty described his young self as a model tourist, faithfully making the rounds of the museums and galleries from Sweden and Russia to Greece and Egypt, but he could recall being impressed by only one painting, the *Venus* by Titian in the Uffizi.

After spending the tense early days of World War I in London with his parents, Mr. Getty began a trial year of joint ventures in the Oklahoma oil fields with his father. In 1916 he brought in his first producing oil well, qualifying at age twenty-three as a millionaire. The roaring excitement of the Oklahoma boom towns may have had an attraction for the quiet and persistent young man, but he returned to Los Angeles in July 1916. When the United States entered World War I, he applied to the United States Army Air Service but was not called. He gradually adopted a pattern of living about seven months of the year in Europe and five months in the United States.

When his father died in May 1930, Mr. Getty became president of George F. Getty Incorporated, the family oil firm. Perhaps as a relief from his business responsibilities, he found his interest in art developing. He read voraciously on the subject and visited museums whenever he could. The United States and Europe were enjoying a period of tremendous prosperity, and great numbers of wealthy collectors—including

Hearst, the Mellons, and the Rothschilds—bid against each other enthusiastically for whatever came on the market. Following the panic of 1929 and the Depression, the time seemed right to begin collecting works of art.

In March 1931 Mr. Getty purchased his first art object of value, a landscape by the seventeenth-century Dutch artist Jan van Goyen, at auction in Berlin for about $1,100. This purchase was followed two years later, in November 1933, by the acquisition at auction in New York of a group of ten paintings by the Spanish Impressionist Joaquin Sorolla y Bastida, whom Mr. Getty admired for his brilliant treatment of sunlight.

Mr. Getty often said that he was inspired to collect the decorative arts after leasing a penthouse in 1936 from Mrs. Frederick Guest on Sutton Place South in New York. The apartment was furnished with French and English eighteenth-century pieces to whose charm Mr. Getty responded enthusiastically. In later years he was frequently asked why he devoted so much of his energy to acquiring carpets, tapestries, and furniture. He claimed that there were two reasons:

> First, I do not subscribe to the theory that only paintings, sculpture, ceramics, and architecture qualify as major fine arts. To my way of thinking, a rug or carpet or a piece of furniture can be as beautiful, possess as much artistic merit, and reflect as much creative genius as a painting or a statue. Second, I firmly believe that beautiful paintings or sculptures should be displayed in surroundings of equal quality. Few men would dream of wearing a fifty-cent necktie with a three-hundred-dollar suit, yet all too many collectors are apparently content to have their first-rate paintings hang in a room with tenth-rate furniture that stands on a floor covered with a cheap machine-loomed carpet.

A period of intense buying began in June 1938, when Mr. Getty purchased many of the major pieces of French eighteenth-century furniture offered at the Mortimer Schiff sale in London. Fear of impending war inhibited bidding, and prices were far below what had been anticipated. Bidding for himself, Mr. Getty acquired at one stroke the beginnings of a significant collection. In later years he said, "My collection like Gaul can be divided into three parts: before Schiff, Schiff, and after Schiff." Among the purchases were a magnificent Savonnerie carpet, the famous roll-top desk by Bernard Molitor (fig. 2), a side table with Sèvres plaques by Martin Carlin, a damask settee and chairs by Jean-Baptiste Tilliard *fils* (see p. 190), and two Chinese porcelain vases with French gilt bronze mounts.

The purchase that may have given Mr. Getty the most pride was the Ardabil Carpet (fig. 3), a large Persian carpet made during the first half of the sixteenth century to adorn the most holy of all the Persian religious shrines, the mosque of Safi-ud-din. When Mr. Getty first saw the carpet on loan from Duveen to an exhibition of Persian art in Paris in 1938, he declared it to be one of the most beautiful things he had ever seen. His

Figure 3. Ardabil Carpet. Persian, Safavid dynasty, ca. 1540. Wool and silk, 729 x 409 cm (23 ft. 11 in. x 13 ft. 5 in.). Los Angeles County Museum of Art 53.30.2.

*Figure 4. Rembrandt van Rijn, Dutch, 1606–
1669. Portrait of Marten Looten, 1632.
Oil on panel, 92.7 x 76.2 cm (36½ x 30 in.).
Los Angeles County Museum of Art 53.50.3.*

*Figure 5. Portrait Bust of the Empress Sabina.
Roman, ca. A.D. 135. Marble, H: 43 cm
(10⅞ in.). 70.AA.100.*

attempts to acquire it immediately were frustrated, as Duveen wished
to keep it for his own private collection. However, in the summer of
1938, Mr. Getty's persistence and the fear of war persuaded Duveen
to sell him the carpet.

In succeeding months Mr. Getty bought a large number of paintings,
the beginning of the Museum's present collection. The acquisition of
Rembrandt's *Portrait of Marten Looten* (fig. 4) was controversial. Mr. Getty
bought the painting, executed by Rembrandt in 1632 when he was
twenty-six, at the sale of the Anton W. W. Mensing collection. Mr. Getty
had been attracted by the painting when it was on display in the Museum
Boymans-van Beuningen, Rotterdam, and bid for it anonymously. When
his bid was successful, there was an outcry in the Netherlands about losing a national treasure to an "unnamed American." Still anonymously,
Getty loaned the portrait for exhibit in the Fine Arts Pavilion of the 1939
New York World's Fair, thus enabling him "to share his joy in owning the
masterpiece with millions of people."

During the summer of 1939 Mr. Getty was in Rome, where he spent
much time looking at ancient Roman monuments and seriously considered purchasing the magnificent Nilotic mosaic at Palestrina, which
then belonged to Prince Barberini. While Mr. Getty did not acquire the
mosaic, he did buy two Roman portraits of the empresses Livia and
Sabina (see fig. 5).

World War II completely curtailed Mr. Getty's activity as a collector.
He threw most of his energies during the war years into expanding and
running Spartan Aircraft Corporation in Tulsa, manufacturing aircraft
and training pilots. Immediately following the war, Mr. Getty remained

in the United States, converting Spartan to peacetime production. But by 1947 he was back in Europe, living in hotel rooms and negotiating deals with the burgeoning European oil market and for the development of oil fields in North Africa and Arabia. Development of a fully integrated worldwide oil network—producing, transporting, refining, and marketing—soon made Mr. Getty's company a strong competitor to the giant Seven Sisters of the oil world.

Mr. Getty's major avocation continued to be centered around art. He visited museums, archaeological sites (see fig. 6), and art dealers, and carefully recorded in his diary works of art that impressed him (and often their prices). By the early 1950's he again had begun to add to his collections. He also began to donate works of art from his collection to various museums. Recipients included the Santa Barbara Museum of Art, Oberlin College, and the San Diego Museum of Art. The most dramatic gift was made in 1953, when he gave the Los Angeles County Museum of Art two of his most prized possessions, the Rembrandt *Portrait of Marten Looten* and the Ardabil Carpet; the gift also included other carpets and tapestries. Later in 1953, colleagues, notably Norris Bramlett, persuaded him to consider establishing a museum in his own name. The ranch Mr. Getty had purchased in Los Angeles in 1945 seemed an ideal location. The trust indenture he executed in late 1953 remains the only document in which he specified how his money was to be used. In that indenture he authorized the creation of a "museum, gallery of art and library" and stated the purpose of the trust very simply as "the diffusion of artistic and general knowledge."

It was in the early fifties that Mr. Getty made his first important purchases of antiquities. In rapid succession he bought the life-size marble Lansdowne Herakles (see p. 26), three sculptures from the Earl of Elgin's collection at Broom Hall (including the Elgin Kore), and the fifth-century–B.C. Cottenham Relief. Greek and Roman sculpture was to be a continuing passion. The Herakles remained a personal favorite until the end of Mr. Getty's life. The myths of the virtuous and beneficent demigod had a definite appeal for the American philanthropist, as did the sculpture's connections with the Roman emperor Hadrian, in the ruins of whose villa it had been found, and with a noble British collection.

Mr. Getty was also particularly interested in French furniture. In 1949 he even published a short book entitled *Europe in the Eighteenth Century,* a distillation of his own studies of European culture and history. In a long chapter on French decorative arts, the illustrations are almost entirely of furniture in his own collection. One of his most notable purchases was made soon after, in 1952: the famous double desk by Bernard van Risenburgh that had belonged to the dukes of Argyll since the late eighteenth century (see p. 176).

The J. Paul Getty Museum was opened to the public in May 1954 for two afternoons a week and by reservation for groups on other days.

Figure 6. J. Paul Getty at Jerash, early 1950's.

Although Mr. Getty continued to buy works of art for his private residences, the best pieces began to appear in the Museum, which also acquired a small staff. The distinguished art historian Wilhelm R. Valentiner, who had been director of the Detroit Institute of Arts and, most recently, of the Los Angeles County Museum of Art, was made director in 1954. He remained for two years. However, neither Dr. Valentiner nor his curator, Paul Wescher, exerted much influence on Mr. Getty's acquisitions. George F. Getty II, Mr. Getty's eldest son, became director shortly after Dr. Valentiner and remained until his death in 1973. Mr. Getty himself then served as director until he died.

In the mid-fifties Mr. Getty tried to build all three of his collections—paintings, decorative arts, and antiquities—with some consistency. During this period he met the famous connoisseur Bernard Berenson and passed much time talking with him and studying art in the library at Villa I Tatti in Settignano, not far from Florence. Under Berenson's influence, Mr. Getty began to take an increasing interest in Italian Renaissance painting and to acquire Italian pictures to add to the Dutch and English paintings he already owned. His feelings on art and the art market at this time are expressed in *Collector's Choice,* an unusual book he wrote with Ethel LeVane in 1955. There he summarizes his approach to collecting: "I buy the things I like—and I like the things I buy—the true collector's guiding philosophy."

Mr. Getty enjoyed research and endeavored to study his acquisitions thoroughly, often employing scholars to help him ascertain the aesthetic values, provenances, and former prices of works of art under consideration. Comfortable in six languages—English, French, Spanish, German,

Figure 7. Titian, Italian, 1477–1576, and his workshop. The Penitent Magdalen, *1530. Oil on canvas, 106 x 91.4 cm (41¾ x 36 in.). 56.PA.1.*

Italian, and Latin—and slightly familiar with Greek, Arabic, and Russian, Mr. Getty had an approach to research that was both wide-ranging and individual. He frequently lamented that he lacked sufficient leisure to follow up all the clues.

In some ways Mr. Getty saw himself in the tradition of other great and eclectic collectors, such as his own older contemporary William Randolph Hearst or the Roman emperor Hadrian. In his diary Mr. Getty compared the architecture and investments represented by Hadrian's Villa and Hearst Castle at San Simeon with his own holdings in art and real estate. He was gratified to own works of art that great connoisseurs of the past had owned and cherished. He was hesitant, however, about his paintings collection, which he evidently expanded mostly in order to broaden his new museum's appeal to visitors. He bought some Impressionist canvases and Italian primitives, and—in an attempt to replace the *Marten Looten* with another star—bought the *Penitent Magdalen* by Titian in 1956 (fig. 7). Although it has since become apparent that the painting is only one of seven versions of the composition and is not well preserved, at the time it had a good scholarly reputation and a high price.

In October 1957 an article in *Fortune* magazine listed Mr. Getty, until then a virtually unknown businessman, as the richest American. From then on the American public began to use the Getty name as a synonym for wealth. Mr. Getty was uncomfortable about the attention paid to his personal fortune and attempted to deflect it into discussions of the number of jobs his corporations created through reinvestment of profits. Also in 1957 a gallery was added to the small ranch museum in Los Angeles, and all of the recent antiquities and decorative arts purchases were installed (figs. 8, 9). Then Mr. Getty abruptly informed Paul Wescher by cable: "I think the JPG museum has enough pictures. We have no space for any more." This resolution was kept for almost a decade.

Much of Mr. Getty's interest for the period between 1958 and 1968 was centered on Sutton Place, a manor house built twenty-five miles south-

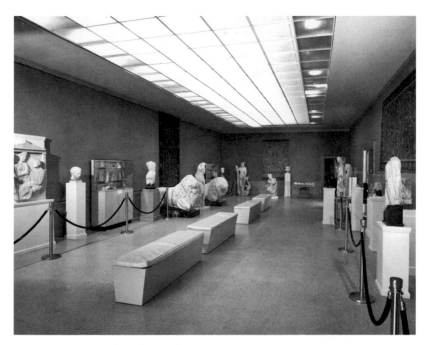

Figure 8. Antiquities gallery, The J. Paul Getty Museum, ca. 1965.

Figure 9. Decorative arts gallery, The J. Paul Getty Museum, ca. 1965.

Figure 10. The yew alley, Sutton Place, 1959. Photo: Julius Shulman.

Figure 11. J. Paul Getty being interviewed by Intertel in the Great Hall, Sutton Place, 1959. Photo: Julius Shulman.

Figure 12. Attributed to Peter Paul Rubens, Flemish, 1577–1640. Diana and Her Nymphs Departing for the Hunt, 1615/18. Oil on canvas, 235 x 182.9 cm (92½ x 72 in.). 71.DA.14.

west of London in 1521–1526 by one of Henry VIII's courtiers. The house was acquired in 1959 by a subsidiary of the Getty Oil Company both as the parent company's new international headquarters and as Mr. Getty's personal residence (see figs. 10, 11). He continued to acquire art through a subsidiary corporation, Art Properties, Inc. One major acquisition, made in 1961, was *Diana and Her Nymphs Departing for the Hunt* (fig. 12), attributed to Rubens, which hung in the Great Hall at Sutton Place until it was given to the Museum in 1971. Another was the Rembrandt portrait of a man with a knife, now identified as *Saint Bartholomew* (see p. 111), in 1962. Both acquisitions were notable for the high prices Mr. Getty was now willing to pay for paintings he liked, although later that year he was appalled by the amount the Metropolitan Museum of Art paid for Rembrandt's *Aristotle Contemplating the Bust of Homer*, which started a new era of high prices.

In the 1960's Mr. Getty asked several of his scholarly friends to help him describe the results of his first thirty years of collecting. Jean Charbonneaux, curator at the Louvre; Julius S. Held, professor at Barnard College, Columbia University; and Pierre Verlet, curator of furniture and objets d'art at the Louvre, helped him write *The Joys of Collecting*, published in 1965. The three scholars attempted to point out both the strength and individuality of the collection. Mr. Getty's own introductory essay summarizes his "addiction" to collecting art and details some of the highlights of his career as a collector. It describes vividly "the romance and zest—the excitement, suspense, thrills, and triumphs—that make art collecting one of the most exhilarating and satisfying of all human endeavors."

In 1964 Mr. Getty bought the full-length *Portrait of a Man* by Veronese (see p. 98) that was subsequently loaned to the National Gallery, London. By 1967 he had resumed buying at auctions. His interest in his Museum also revived, and in the coming years he donated large sums for the purchase of works of art, most notably the *Portrait of Agostino Pallavicini* by Anthony van Dyck (see p. 105).

During 1968 Mr. Getty decided to expand his Museum. Gradually, he evolved the idea of a major art museum that he could leave as a gift to the people of Los Angeles. As he wrote,

> ...there were other, and for me, overriding considerations. It was my intent that the collections should be completely open to the public, free of all charges—be they for admission or even for parking automobiles. Nothing of this sort could be insured if the museum were under the control of a city, state—or even the Federal—government.

To this end he eventually invested more than $17 million in a new building and endowed the Museum to meet operating expenses. Various alternatives were proposed: Spanish Colonial style to suit the existing ranch building, Neoclassical to suit the Greek and Roman sculpture collection.

Then—quite unexpectedly—one evening at Sutton Place Mr. Getty announced to a group of guests that he wished to build a separate large building on the ranch site and that it might be an accurate re-creation of the Villa dei Papiri in Herculaneum. The villa had been one of the largest ever built in the ancient Roman empire and had possibly been owned by Lucius Calpurnius Piso, father-in-law of Julius Caesar and patron of the Epicurean philosopher Philodemus. The villa had stood outside the walls of Herculaneum on the Bay of Naples and had been destroyed and buried along with that city and Pompeii when Mount Vesuvius erupted in A.D. 79. Mr. Getty eagerly reviewed the scanty information about the villa's architecture, the considerable number of sculptures it housed, and the papyrus scrolls retrieved from it through tunnel excavations in the middle of the eighteenth century. He wanted a museum building that would itself be a work of art or of historical interest, and the decision to recreate the Villa dei Papiri must be seen in this light (see figs. 13, 14).

Throughout his life, of course, Mr. Getty had been fascinated by ancient Greece and Rome, and he had two houses in Italy, one near Ostia and one near Naples. The best explanation for his choice of architectural prototype is still in his own words:

> The public should know that what they will finally see wasn't done on a mere whim or chosen by a committee delegated for such a task. It will simply be what I felt a good museum should be, and it will have the character of a building that I would like to visit myself...the principal reason concerns the collection of Greek and Roman art which the museum has managed to acquire...and what could be more logical than to display it in a classical building where it might originally have been seen? There is, I believe, no other place in the world where one can see such a building in any state except ruins, as one sees them now in Pompeii. There are replicas and imitations of ancient public buildings but none of a private structure—so this one should provide a unique experience.

Figure 13. Aerial view of the J. Paul Getty Museum, 1980. Photo: Les Nakashima.

Figure 14. View of the Main Peristyle Garden, The J. Paul Getty Museum, 1980. Photo: Julius Shulman.

Ground was broken in December 1970, and three intense years of work produced the results that delight more than four hundred thousand visitors a year. During construction, which started in mid-1971, Mr. Getty and the trustees realized that the new building would provide an enormous increase in gallery space. From 1970 to 1974, a large number of works of art was purchased, and the professional staff was enlarged. Buying reached a peak in 1971 and 1972, culminating in the purchase at auction of the famous *Death of Actaeon* by Titian—which, however, the Museum was not permitted to export from England and which now belongs to the National Gallery, London.

The new J. Paul Getty Museum building opened its doors to the public in January 1974. Greek and Roman art set the tone because of the architecture and gardens and also because the entire main floor was devoted to antiquities. It was generally recognized that this part of the collection was of very great importance. Relative to other museums, however, the French furniture was the finest of the three Getty collections, with paintings remaining the least developed. Acquisitions continued to be made at a lesser rate in all three areas between January 1974 and June 1976, the month of Mr. Getty's death, but after four years of intensive buying, it was time to shift the Museum's emphasis to assimilation, research, and catalogue preparation.

At his death, Mr. Getty bequeathed a vast legacy to his Museum, leaving it to the trustees' discretion to decide how it should best be used. In the first years after he died, the collections grew significantly through the receipt of Mr. Getty's private collection and some purchases. Significant acquisitions included the late fourth-century-B.C. Greek bronze Victorious Athlete attributed to a follower of Lysippos, for which Mr. Getty had negotiated for several years; the *Saint Andrew* by Masaccio; and the corner cupboard by Jacques Dubois (see pp. 32–33, 92, 178–179).

By April 1982, with the receipt of the proceeds of Mr. Getty's estate, the J. Paul Getty Trust already had begun to prepare for its transformation from a small museum into a visual arts institution of international significance. The trustees made commitments to six operating activities in addition to the further growth of the Getty Museum. These activities were the Getty Center for the History of Art and the Humanities, the Getty Conservation Institute, the Getty Art History Information Program, the Getty Center for Education in the Arts, the Museum Management Institute, and the Program for Art on Film, a joint venture with the Metropolitan Museum of Art. It is hoped that the programs operated by the Getty Trust will make important contributions to the visual arts and that their activities may lead to new forms of collaboration as well as new knowledge in the field. Both the art collections that were shaped directly by Mr. Getty and the new programs of the Trust are tributes to his belief in the power of the visual arts in humanity's past and future.

N.B. Mr. Getty's own observations as quoted here have been taken from the following books written by or with him: *The History of the Oil Business of George F. and J. Paul Getty from 1903–1939*, Los Angeles: n.p., 1941; *Europe in the Eighteenth Century*, Santa Monica, 1949; *Collector's Choice* (with Ethel Le Vane), London: W. H. Allen, 1955; Ralph Hewins, *J. Paul Getty: The Richest American*, London: Sidgwick and Jackson, 1961; *My Life and Fortunes*, New York: Duell, Sloan and Pearce, 1963; *The Joys of Collecting*, New York: Hawthorn Books, 1965; and *As I See It*, Englewood Cliffs: Prentice-Hall, 1976.

ANTIQUITIES

The beginning of the Museum's antiquities collection can be dated to J. Paul Getty's first purchase of an object of classical art, a small terracotta sculpture bought at a Sotheby's auction in London in 1939. The thirty-eight years that passed between that time and Mr. Getty's last direct acquisition, a marble head of a Roman youth, witnessed the creation of a collection of ancient Greek and Roman art which had become the third most important of its kind in the United States by the time of Mr. Getty's death.

For many years the collection's greatest strength was in sculpture, which was the heart of the antiquities collection when Mr. Getty was alive. The marble, bronze, and terracotta representations still provide an exceptional perspective on the artistic achievements of the Greek and Roman periods. Many of the Museum's most interesting examples of ancient sculpture, including the dedicatory group with a portrait of Alexander the Great (see p. 24) and the Tarentine terracotta group of a musician and sirens (see p. 41), were acquired under Mr. Getty's personal direction.

In the years since his death, the collection has continued to grow. Several major acquisitions of Greek sculpture have been made, including the Victorious Athlete (see pp. 32–33) and the brilliantly painted group composed of two griffins attacking a fallen doe (see p. 23). Other recent additions have included a number of Cycladic marble sculptures of the Aegean Bronze Age and the cult statue of a goddess from the Classical period (see pp. 20–21).

Important acquisitions have been made in other areas. The collection of Attic and Italiote vases has grown since Mr. Getty's death to become the department's strongest holding. Included are masterpieces such as the kalpis painted by the Kleophrades Painter, the large drinking cup painted by Onesimos, and the drinking cup painted by Douris (see pp. 45, 46, 47). Other noteworthy acquisitions have been made in the areas of Roman bronze sculpture, Greek and Roman gems, and luxury wares such as Hellenistic silver and Roman glass. While the collection's scope has been extended to include fine examples of Egyptian portrait painting and cast bronze portraits from Northern Europe, its focus has been and will continue to be art of the Classical period.

FEMALE FIGURE

Cypriot, ca. 2500 B.C.

Limestone

H: 39.5 cm (15½ in.)

83.AA.38

Among the earliest sculpted objects of artistic importance from the Aegean and eastern Mediterranean areas, Cypriot female figurines range in height from only an inch or so to over a foot. The Museum's statuette is one of the largest known. With head raised, arms extended, and legs drawn up, she appears to be seated, perhaps giving birth. An ancient break and repair can be seen at her left shoulder. The stylized simplicity of the composition and the barely modulated surfaces of her breasts and beaded ornaments combine to project an image of direct and almost mystical power.

IDOL

Namepiece of the Steiner Master

Early Cycladic, Late Spedos Variety, ca. 2500–2400 B.C.

Island marble

H: 59.9 cm (23⅝ in.)

88.AA.80

Discovered on the Greek islands known as the Cyclades, female idols such as this one probably served a funereal function, since many have been found in burial contexts. The idol pictured here belongs to the best-known and most numerous type, which reclines with folded arms. Made during the early Bronze Age, it shows remarkable sophistication in its execution. The details of the human body are reduced to a minimum, and the figure is a flattened, schematic representation that approaches pure abstraction. The slender, elongated torso is carved with exceptional subtlety and skill. Incision delineates the arms from the body, separates the thighs, and defines the abdomen and pubic triangle. Many idols originally were enhanced with brightly colored pigments that accentuated anatomical features and decorative markings.

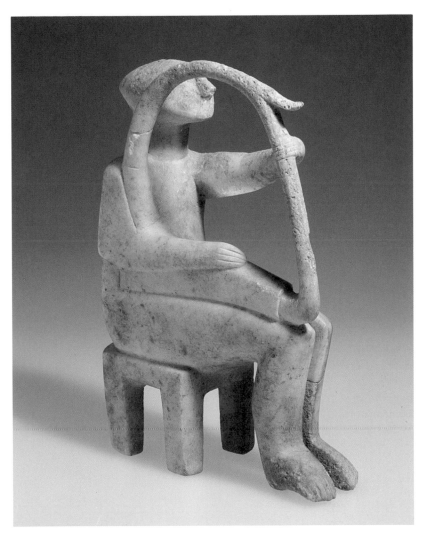

HARPIST
Early Cycladic II, ca. 2500 B.C.
Island marble
35.8 x 9.5 cm (14¹/₁₆ x 3¾ in.)
85.AA.103

Little is known of the prehistoric culture of the Cyclades which produced this beautifully modeled figure of a harpist. He sits on a four-legged stool, his left hand encircling the frame of his instrument and his right hand resting on its sound box. Different pigments were applied to indicate the features of the face and hair, but these have disappeared with time, leaving ghostlike impressions on the surface of the marble. The original purpose of the figure is unknown. Such figures may have been created as protective idols that accompanied their owners to the grave.

GRAVE STELE OF THE HOPLITE POLLIS
Greek, ca. 480 B.C.
Marble
149.8 x 44.5 cm (59 x 17½ in.)
90.AA.129

This grave marker probably stood in or near the Greek city-state of Megara, midway between Athens and Corinth. It shows a hoplite, or lightly armed foot soldier, carrying a spear and round shield. He wears a helmet and a scabbard holding a sword at his left side. Although the face is damaged, the figure is an excellent example of early fifth-century-B.C. Greek sculpture, illustrating the transition from the Archaic to the early Classical style. The monument commemorates a hoplite who fell in battle. The inscription, written in Megarian script, reads: "The beloved son of Asopitos, I, Pollis, speak; not having died a coward . . ."

CULT STATUE OF APHRODITE (?)
South Italian or Sicilian, ca. 400 B.C.
Limestone and Parian marble with polychromy
H: 2.3 m (7½ ft.)
88.AA.76

The over-life-size proportions, quality of execution, and fact that the figure is finished on all sides suggest that this statue was made as a cult image, or sacred representation of the deity, intended to stand within a temple. The voluptuous proportions of the figure and the intentional suggestion of breezes moving the drapery make it most likely that the subject is Aphrodite, the goddess of love. Sculpted from an unusual combination of fine limestone—used for the draped body—and Parian marble—used for the head, preserved arm, and foot—this statue is surely a product of the Greek colonies of South Italy and Sicily, yet it strongly reflects the style of late Classical sculpture from Athens.

GRAVE STELE OF A WARRIOR AND HIS WIFE

Greek, Athens, end of fifth century B.C.
Pentelic marble
102.5 x 43.3 x 15.3 cm
(40⁵/₁₆ x 17¹/₁₆ x 6 in.)
83.AA.378

This stele stood as the marker above a family grave. The names of the deceased, Philoxenos and Philomene, are inscribed near the top. Philoxenos, clearly a soldier, probably fell in battle. The handshake symbolizes his farewell to his wife and their eternal union. Details of the figures and perhaps of the architectural frame would have been added originally in paint.

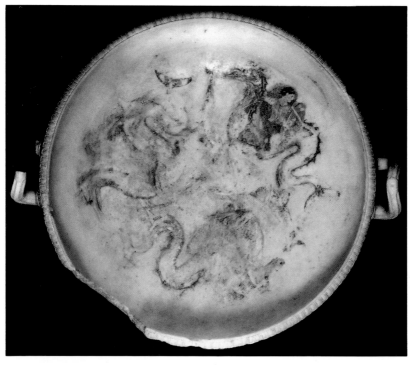

SCULPTURAL GROUP OF TWO GRIFFINS ATTACKING A FALLEN DOE

South Italian, late fourth century B.C.
Marble with polychromy
95 x 148 cm (37⅜ x 58¼ in.)
85.AA.106

The surviving pigments on this sculptural group provide a vivid reminder that all ancient marbles were once brightly painted. In a remarkable feat of sculpting, the artist has captured the notion of dangerous beauty by juxtaposing the lithe, sinuous forms of the griffins with the brutality of their attack on the ill-fated doe. The symmetrical, upright, sickle-shaped wings were functionally necessary, for the channels carved between the wings indicate that the piece served as a support, probably for a ceremonial table.

LEKANIS

South Italian, late fourth century B.C.
Marble
30.8 x 70 cm (12⅛ x 27⁹⁄₁₆ in.)
85.AA.107

The decoration on the interior of this unique ceremonial basin provides a sense of the original appearance of Greek wall painting. Rendered in vivid polychromy, Thetis and two Nereids ride to the left, one on a spirited hippocamp, the other two on keti, or sea monsters. Each carries a piece of armor for Thetis' son Achilles, the greatest Greek hero in the Trojan War. This conceit would have been especially charming when the basin was filled with water, as the women would have appeared to be riding across the sea.

HEAD OF ALEXANDER

Greek, end of fourth century B.C.
Marble
H: 28 cm (11 in.)
73.AA.27

Identifiable by the manelike, swept back hair and the deep-set, upturned eyes, this head is undoubtedly a portrait of Alexander of Macedon. This portrait type of the youthful conqueror was created by the fourth-century artist Lysippos, the only sculptor Alexander allowed to portray him during his short life. This head and that of a youth which accompanies it and which may represent Hephaistion, Alexander's beloved companion, belonged to a sculptural group depicting Alexander and other figures that was made sometime after his death in 323 B.C.

PORTRAIT OF CALIGULA

Roman, ca. A.D. 40
Marble
H: 43 cm (16¹⁵⁄₁₆ in.)
72.AA.155

Gaius Julius Caesar Germanicus, the third emperor of Rome (r. A.D. 37–41), spent his early childhood in a Roman army camp in Germany. Since his parents dressed him in a military costume that included *caligae*, the footgear of a soldier, the legionnaires gave him the nickname Caligula, or Little Boots. He became emperor of Rome at the age of twenty-five but was assassinated within four years because his reign had turned despotic and his behavior, cruelly erratic. Because of his unpopularity, most portraits of Caligula were destroyed after his death. This surviving likeness reflects late Julio-Claudian classicism, a style popular during the reign of Claudius (A.D. 41–58), Caligula's successor.

STATUE OF A CENTAUR

Roman, ca. A.D. 90

(Centaur) rosso antico; (base) breccia

H: (centaur) 157 cm (61¹³/₁₆ in.); (base) 86.6 cm (34½ in.)

82.AA.78

This statue of a young centaur, preserved on its original base, is carved in rosso antico, or red marble, which even in antiquity was considered a precious material and expresses the extravagant taste of the Roman ruling class. A sumptuous example of decorative sculpture of the time, this centaur originally stood as part of a group in a palace of the emperor Domitian (r. A.D. 81–96) near modern-day Castel Gandolfo in Italy. The mythical half-man/half-horse has a pigskin draped over his left arm; he grins broadly and holds a rustic syrinx, or panpipe, in his other hand. As woodland creatures with a reputation for being lustful and overly fond of wine, centaurs often took part in the revels of Bacchus.

LANSDOWNE HERAKLES
Roman, ca. A.D. 135
Pentelic marble
H: 193.5 cm (76³/₁₆ in.)
70.AA.109

Found at Tivoli in 1790 or 1791 in the ruins of the villa of the emperor Hadrian
(r. A.D. 117–138) and until 1951 in the collection of the Marquess of Lansdowne,
the Lansdowne Herakles was one of J. Paul Getty's favorite pieces. Although
related in appearance to works attributed to fourth-century-B.C. sculptors such
as Skopas and Lysippos, the statue has an eclectic style that is purely Roman.
The young Herakles is shown larger than life-size, holding the club with
which he slew the Nemean lion. As an exemplar of human achievement and
the subject of philosophical discussion, he was an appropriate model for a
Roman emperor.

TORSO OF A PUGILIST, PERHAPS HERAKLES

Roman, Alexandria (?),
second century A.D.
Parian marble
H: 58 cm (22¹³/₁₆ in.)
83.AA.11

Though made as a separate piece, this torso was part of a sculpture that showed the subject engaged in active combat. The vigorous pose and carefully articulated musculature reflect the interests and innovations of late Hellenistic sculpture, but the incised pupils betray the bust's Roman Imperial date. The features recall those most often associated with Herakles, but positive identification is impossible.

FRONT PANEL OF A SARCOPHAGUS

Roman, ca. A.D. 210
Marble
54.5 x 214 cm (21½ x 84¼ in.)
76.AA.8

The scene on this sarcophagus illustrates the story of Selene, the goddess of the moon, and her lover, Endymion. Jealous of their happiness, Zeus caused Endymion to fall into eternal sleep. Selene, however, continued to visit him nightly in a cave on Mount Latmos. Near the center of this panel, Selene, accompanied by three small Erotes, dismounts from the chariot with which she draws the moon across the sky. Endymion lies asleep on the ground; behind him Hypnos, the god of sleep, pours a magic potion to ensure that he will not awaken. To the right Selene appears again as she leaves in her chariot.

PORTRAIT BUST OF AN ANTONINE LADY
Roman, ca. A.D. 150
Carrara marble
H: 67.5 cm (26⁹/₁₆ in.)
83.AA.44

The face of this elegant lady seems to convey a placid yet intriguingly magnetic character. Although her face is unlined, she is no longer young; one detects the confident self-assurance of a Roman matron secure in her status. The hairstyle, once favored by Empress Faustina, and the high polish date the piece to the Antonine period. This bust's excellent preservation, deep psychological insight, and skilled carving make it one of the finest Roman portraits in the Museum.

PORTRAIT OF A MAN
Roman, fifth century A.D.
Marble
H: 25.5 cm (10¹/₁₆ in.)
85.AA.112

This beautifully carved and well-preserved portrait head depicts a balding elderly man. The individualized features suggest that he may have been a person of some importance, perhaps a magistrate or educated man. The fact that the striking blue-gray marble used for the portrait comes from the area around the island of Marmara in modern-day Turkey may indicate that the subject was a Roman citizen of Asia Minor. The unknown artist skillfully utilized the contrasting techniques of deep drill work and high polish to differentiate between the textures of the subject's smooth skin and tightly curled hair.

FURNITURE SUPPORT REPRESENTING A WINGED FELINE

Tartessian, seventh–early sixth century B.C.
Bronze
H: 61 cm (24 in.)
79.AB.140

Although winged felines were popular in Greek, Phoenician, and Assyrian art, this creature's articulated wings and curved brow point to an origin in Tartessos, the Phoenician colony on the southwest coast of the Iberian peninsula. In the eighth and seventh centuries B.C., native craftsmen were influenced by artistic currents from Phoenician settlements nearby. The struts behind the head and below the front paws suggest that this hollow bronze functioned as the front leg of a throne or other piece of furniture.

FRAGMENTARY SHIELD BAND WITH RELIEFS

Signed by Aristodamos of Argos as maker
Greek, ca. 580 B.C.
Bronze
16.2 x 8 cm (6⅜ x 3⅛ in.)
84.AC.11

The size of this small fragment belies its significance in the history of Greek art, for written across the background of the lowest panel is the earliest recognized signature of a Greek metalworker. This relief provides a valuable document for the identification of an artistic personality and the definition of an important regional style. In the upper of the two panels that preserve figural decoration, a female figure is being led at swordpoint by a male in armor. The presence of the goddess Athena at the right edge of the scene suggests that the subject is the recovery of Helen by her husband, Menelaus. Inscriptions on the lower panel identify Herakles' bride Deianeira on the back of the centaur Nessos, who abducted her.

STATUETTE OF A KOUROS

Etruscan, first quarter of fifth
century B.C.
Bronze
H: 22.5 cm (8⅞ in.)
85.AB.104

This statuette of a kouros, or young
male, is based on a type developed in
Greece during the sixth century B.C.
Rather than retaining the heroic
nudity of the Greek statues, however,
this Etruscan sculptor clothed his
figure in a *tebenna*, the precursor of the
Roman toga. Wrapped around his
right hip, one end of the long garment
is pulled over his left shoulder from
behind, and the other end is thrown
over his left forearm. The decorative
patterns at the edges of the cloth, the
diagonals of the upper border, and the
dots along the lower edge represent
elaborate embroidery. These patterns
and the individual strands of hair were
incised after casting.

STATUETTE OF A
BEARDED MAN

Etruscan, Piombino (?), ca. 460 B.C.
Bronze
H: 17.5 cm (6⅞ in.)
55.AB.12

Said to have been found in Piombino,
this figure is a fine example of Severe-
style Etruscan statuary. The rigid
frontality of its pose and the advanced
left leg are vestiges of the earlier
Archaic style, but the more naturalistic treatment of the torso and face
are new. The figure wears a *tebenna*.
Perhaps the now-missing object he
held in his clenched left hand was a
scepter, justifying his frequent identification as Tinia, the Etruscan
equivalent of the king of the Greek
gods, Zeus.

STATUETTE OF A DEAD OR SLEEPING YOUTH

Greek, ca. 480–460 B.C.

Bronze inlaid with copper

13.5 x 7.3 cm (5⁵⁄₁₆ x 2⁷⁄₈ in.)

86.AB.530

The beautifully articulated details of this figure are the work of a master crafts-man. Copper inlay was used to enhance parts of the anatomy. The objects once held in the figure's hands are now missing. A hole on the back of his right shoulder and small, regular areas of corrosion indicate that he may have been supported by another figure or placed within a landscape setting.

STATUETTE OF A KNEELING SATYR

Greek, ca. 480–460 B.C.

Bronze

H: 10 cm (3⁷⁄₈ in.)

88.AB.72

Mythological creatures who were half-horse and half-human, satyrs represented the embodiment of uncivilized nature, and their appetites for wine and sex were notoriously uncontrolled. No exception, this diminutive figure drinks from a large *keras*, a drinking horn used for undiluted wine, and his complex pose, with limbs extended in every direction, does not disguise the fact that he is ithyphallic. Though the bestial head reflects the late Archaic tradition, the innovative posture and advanced anatomical accuracy of the torso place the figure firmly within the transitional period between Archaic and Classical art.

TRAPPINGS FOR A HORSE: HEAD PIECE
(*Prometopidion*) AND CHEST GUARD

South Italian, end of sixth century B.C.

Bronze; ivory and amber inlay

(*Prometopidion* [original]) 45 x 17.2 cm (17¹¹/₁₆ x 6¾ in.); (chest guard [restored]) approx. 107 x 25 cm (42⅛ x 9⅞ in.)

83.AC.7

Both the precious materials and the beautiful decoration of these trappings distinguish them as ceremonial armor intended, perhaps, for a champion chariot team. The elaborately detailed head piece takes the shape of the face of a warrior wearing a helmet with ram's-head cheek pieces. The warrior's eyes are inlaid with ivory and amber. The chest guard is decorated with an incised frontal quadriga flanked by flying Nikai, or Victories, who carry wreaths.

STATUE OF A VICTORIOUS ATHLETE

Greek, late fourth century B.C.

Bronze

H: 151.5 cm (59⅝ in.)

77.AB.30

Few monumental Greek bronze sculptures have survived. Among them, this representation of a victorious young athlete is one of the finest examples from the last decades of the fourth century. Standing confidently, the subject raises his right hand beside his head, on which he has just placed the olive wreath of victory. Each city or sanctuary that hosted athletic games awarded wreaths made from a specific plant to their winners, and the olive wreath was the prize given to the victor in the most prestigious of all of these contests, at Olympia. This sculpture most likely was dedicated there.

Although the artist is unknown, the style of the sculpture strongly reflects the innovations of Lysippos, the fourth-century-B.C. sculptor who specialized in the casting of bronze figures. Among the features that may be termed Lysippan are the open stance, the head's small size relative to the body, and the suggestion of torsion in the pose, which encourages the observer to walk around the piece.

HYDRIA

Greek, mid–fourth century B.C.
Bronze
48 x 39.5 x diam. (mouth) 17 cm
(18⅞ x 15⅝ x 6¹¹/₁₆ in.)
79.AC.119

The hydria was used in antiquity as a
water jar. The horizontal side handles
were convenient for carrying and the
vertical handle, for pouring. This
example is decorated with floral patterns at the mouth, on the foot, and at
the base of the fluted handles. The
separately hammered relief plate below
the vertical handle shows Herakles
carrying his club in one hand and the
baby Eros in the other.

PORTRAIT OF A MAN

Roman, Asia Minor (?), first
century B.C.
Bronze
H: 29.5 cm (11¼ in.)
73.AB.8

A literary description and the striking
similarity between this head and certain early Roman coin portraits have
resulted in the tentative identification
of the subject as Lucius Cornelius
Sulla, the celebrated Roman general
of the late second century B.C. It
is the earliest Roman portrait in the
Museum's collection and is a masterpiece of technical achievement in
casting. The eyes once were inlaid in
another material.

BUST OF A FEMALE

Roman, late first century B.C.-early
first century A.D.
Bronze
H: 16.5 cm (7½ in.)
84.AB.59

Remarkably detailed and well preserved, this small bust of a Roman
lady attests to the sensitivity of the
craftsman responsible for its manufacture. The head was carefully worked
after casting to articulate the details
of the intricately braided and knotted
hair. The earlobes are drilled for earrings (now lost), and the inlaid eyes
of glass paste, which usually have
not survived from ancient statues,
are preserved here and lend a
striking appearance to the face.
The bust, set atop a round base,
probably was placed in a household shrine.

PORTRAIT OF MENANDER

Roman, late first century B.C.–
early first century A.D.
Bronze
H: 17 cm (6¹¹/₁₆ in.)
72.AB.108

Prior to the publication of this small
bust, the subject's identity had been
suggested but was unproven. The
Greek inscription on the base confirms
the tentative identification by labeling
the figure as Menander, the popular
fourth-century-B.C. playwright of
entertaining comedies. This bust is
based on an early third-century Greek
statue. The outstanding craftsmanship
is particularly evident in the treatment
of the hair and modeling of the face.

THYMIATERIA
Roman, first half of first century A.D.
Bronze inlaid with silver
23.2 x 13.3 cm (9⅛ x 6 in.); 19 x 9.5 cm (8 x 3¾ in.)
87.AC.143–.144

These statuettes of a singer and a comic actor sitting atop altars were used as incense burners. The singer's identity is based on the sistrum—a musical instrument that functions as a rattle—that he holds in his right hand. The actor wears the clothing and mask associated with productions of Greek New Comedy. The mask's wide open mouth was appropriate for comic effect and the amplification of speech. Both figures are hollow, and the altars on which they sit are pierced underneath for ventilation. The perfumed smoke of the incense burned inside would have emerged through their open mouths.

PORTRAIT BUSTS
Roman, early first century A.D.
Bronze
H: 40.6 cm (16 in.); 40 cm (15⅞ in.)
89.AB.67.1–.2

This pair of matching bronze portraits is a unique survival from the Augustan period. The striking style and summary treatment of the backs of the heads indicate an origin in the Empire north of Italy, perhaps in the province of Gaul. The two youths, probably brothers, are similar in appearance, but subtle differences in the faces suggest a difference in age. Although the masklike portraits are abstract and idealized, the facial features and hairstyles are characteristic of portraits of members of the family of Augustus (r. 27 B.C.–A.D. 14). The busts were probably placed in a sanctuary in Northern Europe as part of the imperial cult Augustus encouraged.

STATUETTE OF ROMA OR
VIRTUS

Roman, ca. A.D. 40–68
Bronze
H: 33.1 cm (13 ⅛ in.)
84.AB.671

The impact of Greek classicism on
Roman Imperial art is evident in
this fine statuette of a female deity.
The figurine once held objects that
established her identity. She might
represent Roma, the personification
of the city and empire of Rome, or
Virtus, the embodiment of valor.
Common to both are helmets, spears,
and short chitons, or dresses, that
leave the right breast bared. The dis-
tinguishing attribute was displayed
in her right hand: a winged Victory,
if Roma, or a sword, if *Virtus*. This
statuette may have been part of
a multifigured composition that
included bronze statuettes of a goddess and two togate figures.

STATUETTE OF A
GODDESS

Roman, first half of first century A.D.
Bronze
H: 32 cm (12 ⅝ in.)
84.AB.670

Clad in a long tunic and mantle, this
figure is identified as a deity by her
diadem. Her original attributes are
missing, but the positioning of her
hands suggests that she held a patera,
or offering dish, in her right one and a
scepter or spear in her left. Numerous
Roman deities and personifications
are represented with these attributes
on coins of the period. This deity once
was part of a decorative composition.
A hole in the back may have been used
to secure the statuette to a chariot or
piece of furniture.

RELIEF WITH TWO TOGATE MAGISTRATES
Roman, second quarter of first century A.D.
Bronze
26 x 13.8 cm (10¼ x 5⁷⁄₁₆ in.)
85.AB.109

These figures, along with the Museum's statuettes of Roma (or *Virtus*) and a
goddess (see p. 38), probably were part of a composition that may have decorated a ceremonial chariot or piece of furniture. Both figures turned their
attention to some action taking place to the left; a break along the right edge
suggests that the composition continued on that side, too. The draping and
fastening of the garments and the hairstyles are characteristic of bronzes made
during Nero's reign (A.D. 54–68). The older man is most likely a priest; his scroll
may contain ritual formulas appropriate to the ceremony performed in the missing section.

THYMIATERION

South Italian, ca. 500–480 B.C.

Terracotta

44.6 x diam. (incense cup) 6.9 cm
(17 ⁹⁄₁₆ x 2 ³⁄₄ in.)

86.AD.681

Nike, the winged goddess of victory,
carries on her head a shallow incense
cup, its openwork lid surmounted by a
finial in the form of a dove. Like the
Archaic korai, or maidens, Nike
stands with one leg advanced and
holds the folds of her skirt in her left
hand. Abundant traces of pink, red,
and blue pigment survive on the
surface of the object, attesting to
its original colorful appearance.
Since it shows no sign of use, this
thymiaterion may have been placed
in a tomb as a grave offering.

PAIR OF ALTARS

South Italian, ca. 400–375 B.C.

Terracotta

41.8 x 34.2 x 29.8 cm (16½ x 13¹¹⁄₁₆ x 11¾ in.); 41.8 x 34 x 28.8 cm
(16½ x 13⅛ x 11⁵⁄₁₆ in.)

86.AD.598.1–.2

These small altars may have served in a private shrine for the presentation of
offerings or incense. The front panels are decorated with groups of figures
in low relief. They appear to depict related parts of a single story about the
impending death of Adonis, god of vegetation. On the left altar three female
attendants rush to the right. On the right altar the dying Adonis sits in a
rocky landscape, supported by his lover, Aphrodite, and attended by two
grieving women.

SEATED MUSICIAN FLANKED BY TWO SIRENS

South Italian, near Taranto, early fourth century B.C.
Painted terracotta
H: (musician) approx. 104 cm (41 in.); (sirens) approx. 140 cm (55⅛ in.)
76.AD.11

This unique trio has often been identified as Orpheus and the Sirens. According to classical mythology, the sirens were demonic singers who lured sailors to destruction with their beautiful songs. Orpheus, the most famous of all mortal musicians in Greek mythology, encountered the sirens on his voyage with Jason and the Argonauts to capture the Golden Fleece. In order to save the crew from certain self-destruction, Orpheus sang to the sirens as the ship sailed past them, causing them to fall silent and, captivated, forget their own songs.

   Although the fact that the seated male figure sings and plays a lyre has suggested his identification as Orpheus, his attire does not match that found in many fourth-century-B.C. South Italian representations of the Greek musician. On contemporary Apulian vases he is often shown in an elaborately embroidered Oriental costume, usually a long flowing robe with a short, capelike chlamys around the shoulders, and a soft Phrygian cap. Since the sirens often appeared during the fourth century in funerary contexts as mourners or as the muses of the underworld, this musician may be a simple mortal imitating Orpheus to insure his own safe passage beyond the fatal song of the sirens.

OINOCHOE
East Greek, Rhodes, ca. 625 B.C.
Terracotta
35.7 x diam. (body) 26.5 cm
(14 x 10⁷/₁₆ in.)
81.AE.83

This trefoil oinochoe, or pitcher with three-lobed spout, is decorated in a black-on-white style of vase painting characteristic of the Orientalizing period of the seventh century B.C. This particular type of East Greek painting is known as the Wild Goat style after its most common motif. Here the figural subjects are exclusively animals, real and imaginary, with floral and geometric patterns used as filler ornaments and framing borders.

BLACK-FIGURED OLPE
Attributed to the Painter of
Vatican 73
Greek, Corinth, ca. 650–625 B.C.
Terracotta
33 x diam. (body)
17 cm (13 x 6⅝ in.)
85.AE.89

The body of this elegant pouring vessel forms one continuous curve from its low foot to its broad rim. The vase is decorated in the manner most frequently used on Corinthian vases, a combination of floral and animal designs. White dot-rosettes enhance each of the two rotellae that flank the tall handle and encircle the neck. Below the red band of the collar are four registers of animal processions. Around the animals are black dot-rosettes, and a ray pattern encircles the lower body above the foot.

ROUND-BODIED HEAD PYXIS
Attributed to the circle of the Chimaera Painter
Greek, Corinth, ca. 570 B.C.
Terracotta
21.8 x diam. 22.2 cm (8⁹/₁₆ x 8¾ in.)
88.AE.105

The body of this large, globular container is decorated with a frieze of real and fantastic creatures rendered in black-figure technique with generous amounts of added red. The creatures include lions, a goat, a bull, and a bearded male siren. Scattered in the field around them are the rosettes and other space-filling ornaments typical of Middle Corinthian vase painting. On the shoulder is a lotus-palmette chain. The handles have been fashioned as busts of female figures wearing red dresses and necklaces. The pyxis was used in antiquity to hold precious oils and powders employed as unguents and cosmetics.

BLACK-FIGURED KYLIX
Attributed to the Boread Painter
Greek, Sparta, ca. 570–565 B.C.
Terracotta
12 x diam. (bowl) 14 cm (4¾ x 5½ in.)
85.AE.121

In the relatively brief period during which vase painting flourished in Sparta, the Boread Painter emerged as one of the more innovative and accomplished Lakonian craftsmen. Although creating balanced compositions presented a challenge for ancient vase painters, here the Boread Painter grouped his figures in a triangular arrangement, successfully filling the circular frame in the interior of the cup. The Corinthian hero Bellerophon attacks the mythical Chimera, a fire-breathing lion with a goat's head on its back and a snake as its tail. As the hero thrusts his spear into the belly of the monster with his right hand, he restrains his winged steed, Pegasus, with his left.

BLACK-FIGURED HYDRIA

Attributed to the Eagle Painter
Caeretan, ca. 525 B.C.
Terracotta
44.6 x diam. 33.4 cm (17 ½ x 13⅛ in.)
83.AE.346

Although the Labors of Herakles were popular subjects in sixth-century-B.C. vase painting, the slaying of the many-headed hydra of Lerna was not a very common representation. The hero had to kill the vicious sea monster by cutting off each head and cauterizing the neck so that new heads could not grow back. Here Herakles' nephew Iolaos assists him. This vase was likely made in Italy by an artist who learned his craft in Ionia. He had little feeling for the architectonic quality of the vase shape and as much interest in the ornamental motifs as in the human subjects.

BLACK-FIGURED ZONE CUP, TYPE A

Attributed to Andokides as potter; attributed to the manner of the Lysippides Painter
Greek, Athens, ca. 520 B.C.
Terracotta
13.6 x diam. 36.4 cm (5⁵/₁₆ x 14⁵/₁₆ in.)
87.AE.22

In the interior of this cup, six revelers recline beneath intertwined grapevines around a central gorgon's head. The exterior is decorated with paired eyes. Between them on one side, Herakles stands before Dionysos, the god of wine, and on the other, Herakles wrestles with the sea monster Triton. The broken rim of the cup was repaired in antiquity with a fragment from another cup.

RED-FIGURED KYLIX, TYPE B

Attributed to the Carpenter Painter
Greek, Athens, ca. 510–500 B.C.
Terracotta
11 x 38.1 cm (4⁵⁄₁₆ x 15 in.)
85.AE.25

Intended for banquet use, drinking cups frequently were decorated with scenes illustrating daily activities such as athletics, feasting, entertainment, and amorous adventures. In the interior of this kylix, an older man and a youth embrace. Around the exterior, athletes practice the discus throw, broad jump, and javelin throw.

RED-FIGURED KALPIS

Attributed to the Kleophrades Painter
Greek, Athens, ca. 480 B.C.
Terracotta
H: (with restored foot and mouth) 39 cm (15⅜ in.)
85.AE.316

The eminent vase scholar Sir John Beazley called the Kleophrades Painter "the greatest pot painter of the late archaic period." The unusual scene on the shoulder of this kalpis, or water jug, is a fine example of his mature style. Although the curving surface presented a challenge, the Kleophrades Painter masterfully adapted his composition to fit the irregular field. The subject is an episode from the adventures of Jason and the Argonauts. Phineus, the blind prophet-king of Thrace, sits before a table, trying desperately to defend himself from three harpies. Because he had misused his gift of prophecy, the gods had punished Phineus by sending these winged demons to steal and befoul his food. The Argonauts, who found him nearly starved to death, saved his life in exchange for his predictions about the future course of their voyage.

RED-FIGURED KYLIX TYPE C
Signed by Euphronios as potter; attributed to Onesimos as painter
Greek, Athens, ca. 490 B.C.
Terracotta
Diam. (bowl [restored]) 46.6 x diam. (foot) 20.5 cm (18⅜ x 8 in.)
83.AE.362; 84.AE.80; 85.AE.385

Remarkable for both its potting and its decoration, this fragmentary late
Archaic kylix represents Attic red-figure vase painting at its best. The scenes
depicted inside and out are episodes from the Trojan War, a subject appropriate
in its grandeur to the monumental proportions of the drinking cup and the
elaborate figural compositions. Inside, both the tondo and the surrounding
zone are filled with scenes from the Sack of Troy. Especially dramatic is the
representation of Neoptolemos' attack on the aged king of Troy, Priam, who
has taken refuge on the altar of Zeus Herkaios. The warrior's weapon is the
body of Priam's grandson, Astyanax. The artist has provided an accurate
and literate account of the tumultuous events surrounding the city's fall.

RED-FIGURED KYLIX

Attributed to the Brygos Painter
Greek, Athens, ca. 490 B.C.
Terracotta
H: (restored) 11.35 x diam. 31.6 cm
(4½ x 12⁷⁄₁₆ in.)
86.AE.286

Ajax, the greatest of the Greek heroes
at Troy after Achilles, rescued the
latter's body from the midst of battle.
Angered and humiliated at losing
Achilles' armor to Odysseus by vote of their comrades, Ajax decided on revenge
but then took his own life instead. Here the painter has depicted the tragic
result, as Tekmessa shrouds the body of Ajax, her lover, who lies fallen on his
suicide weapon.

RED-FIGURED KYLIX

Attributed to Python as potter; signed by Douris as painter
Greek, Athens, ca. 480 B.C.
Terracotta
13.3 x 40.7 cm (5³⁄₁₆ x 16 in.)
84.AE.569

In the tondo a stately bearded figure, perhaps Kekrops, a legendary king of
Athens, sits before an altar holding a kylix that a youthful attendant fills with
wine from an oinochoe, or pitcher. On one side of the exterior, the goddess
Eos pursues Kephalos as Kekrops and Pandion, another mythological king of
Athens, and a third king, perhaps Erechtheus, look on. On the other side Zeus
seizes the Trojan prince Ganymede.

WHITE-GROUND LEKYTHOS

Attributed to Douris as painter
Greek, Athens, ca. 500 B.C.
Terracotta
33.5 x diam. 12.6 cm (13³/₁₆ x 5 in.)
84.AE.770

In a calm, evenly disposed composition around the body of the vase, two young Athenian aristocrats arm themselves in the presence of a boy and a woman. Dressed in a short chiton, one of the youths puts on his greaves, while the other holds his helmet and shield. In the field between the figures is a *kalos* inscription praising the beauty of two youths, Nikodromos and Panaitios.

This type of vase, produced in Athens, was designed to hold precious oils. Such elements as the slender neck and sharply rimmed mouth were ideally suited for pouring the costly liquids. The application of a white slip to the outside of the red clay body enabled the artist to draw his figures with a very precise, clean outline while using a dilute glaze to delineate the interior lines of muscles and drapery folds.

RED-FIGURED MASK KANTHAROS

Attributed to the Foundry Painter
Greek, Athens, ca. 480–470 B.C.
Terracotta
H: (to top of handles) 21.1 x
(to rim) 14.7 x diam. 17.4 cm
(8⁵/₁₆ x 5³/₄ x 6¹³/₁₆ in.)
85.AE.263

Restored from a number of fragments, this cup has two fully modeled masks attached on either side. One is the face of Dionysos, the god of wine, the other, the face of a satyr. Both are appropriate decorations for a cup used for wine. The bowl is decorated with painted figures of youthful athletes, each standing on a short meander band. Elaborate palmettes and lotus buds decorate the handle roots and calyx; the concave rim is surrounded by an ivy garland.

RED-FIGURED VOLUTE KRATER AND STAND

Attributed to the Meleager Painter

Greek, Athens, ca. 390 B.C.

Terracotta

(Krater) 54.2 x 40.6 cm (21¼ x 13⅜ in.); (stand) 16.4 x 34 cm (6½ x 13⅜ in.)

87.AE.93

This krater offers a remarkable combination of painted, gilded, and molded ornament. The subject on the front of the neck is the young Adonis lying on his couch between Aphrodite and Persephone and their attendants. On the other side are three pairs of reclining male figures. These painted scenes and the surrounding floral patterns are enhanced with gilding. Gilt relief heads appear in the volutes of the handles, and fully modeled heads can be seen at the roots of the handles. Echoing the scene on the neck, the stand shows a reclining Dionysos attended by satyrs and maenads. The stem between the stand and vase is a modern reconstruction.

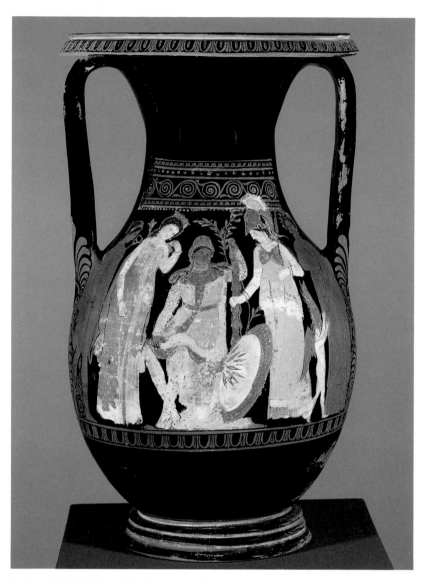

RED-FIGURED KERCH PELIKE
Attributed to the circle of the Marsyas Painter
Greek, Athens, ca. 350 B.C.
48.3 x diam. (mouth) 28.1 x diam. (body) 27.2 cm (19 x 11 x 10¹¹/₁₆ in.)
83.AE.10

Vases in the Kerch style take their name from an area on the Black Sea where numerous examples have been found. The style is characterized by an elaborate and often flamboyant use of polychromy, gilding, and relief work to augment the simple red-on-black scheme of earlier Attic vases. The scene represented on the front of the vase is the Judgment of Paris, the young prince of Troy who had to choose the most beautiful of three goddesses. He sits at the center with the contestants around him.

PANATHENAIC AMPHORA
Greek, Athens, ca. 340/339 B.C.
Terracotta
101 x diam. (mouth) 23.5 x diam. (body) 39.2 cm (39¾ x 9¼ x 15⁷⁄₁₆ in.)
79.AE.147

This particular shape of lidded storage vessel, known as a Panathenaic amphora, was designed for a special purpose. The amphora contained the olive oil gathered from sacred olive groves to be awarded to victors in the quadrennial Panathenaic games. These games, which included both athletic and artistic competitions, were held to honor Athena, the patroness of the city. The painted decoration of the vases, always black-figure, was as traditional as the characteristic shape. On one side the striding figure of Athena was depicted and on the other, an image of one of the competitions. On this vase an elegant team of four horses is shown pulling a chariot that carries a driver and an armed figure known as an *apobates*. The *apobates* would jump on and off the chariot as it raced along in competition with other teams.

RED-FIGURED VOLUTE KRATER
Attributed to the Sisyphus Group
South Italian, Apulia, ca. 430–420 B.C.
Terracotta
H: 63.3 cm (24⅞ in.)
85.AE.102

The pottery produced in the Greek colonies of South Italy preserves a wide
variety of mythological subjects. A scene previously unknown in ancient art
from the myth of Andromeda is the main subject of this krater. Although most
representations show her already bound to tree trunks or a rock while the hero
Perseus slays the sea monster, here she is shown at the beginning of her ordeal.
A youth at the left ties her between tree stumps, while at the right Perseus and
her father, King Kepheus, make a pact that Andromeda will become the hero's
bride if he slays the monster.

RED-FIGURED LOUTROPHOROS

Attributed to the Painter of Louvre MNB 1148
South Italian, Apulia, late fourth century B.C.
Terracotta
90.1 x diam. 35.2 cm (35½ x 13⅞ in.)
86.AE.680

In one of his guises adopted while seducing mortals, Zeus appears here as a
white swan leaping into Leda's embrace, while Hypnos, a personification of
sleep, showers them with drowsiness. To the right and left are Leda's compan-
ions. In the Ionic building above are Zeus and Aphrodite with Eros on her arm.
We might imagine that Zeus has asked for Aphrodite's help in seducing Leda
and that Eros is helping to plead his case. Other figures—Astrape, a personification of lightning; Eniautos, the personification of the calendar year; and
Eleusis, a personification of an important sanctuary—are not clearly connected
to the story.

RED-FIGURED CALYX KRATER
Signed by Asteas as painter
South Italian, Paestum, ca. 340 B.C.
Terracotta
H: 71 cm (28 in.)
81.AE.78

This impressive krater is the largest of eleven known vases signed by Asteas, one of only two artists from the city of Paestum who signed his work. Standing on an elaborately modeled and decorated foot, the vase carries the sole Paestan representation of one of the most popular myths in antiquity: the love of Zeus for the mortal girl Europa. Framed in a unique pentagonal panel, the white bull into which Zeus transformed himself carries Europa across the sea. Pothos, the personification of passionate longing, flies overhead. Various gods, including Aphrodite and Eros, watch the scene from the upper corners like an audience. This type of composition follows a Sicilian tradition with theatrical associations.

ALABASTRON

Hellenistic, Egypt,
second century B.C.
Faience
H: 23 cm (9 ¹/₁₆ in.)
88.AI.135

This perfume bottle's elegant shape
originated in Egypt, and it is named
for the translucent stone from which
the earliest examples were made.
Greek craftsmen borrowed this form,
producing a range of ceramic and
metal variations. In this unique example the artisan has used contrasting
colors of the glassy faience to create
decorative patterns reflecting those
found on Hellenistic silver vessels.
This alabastron was made in three sections: the mouth, bottom, and body.

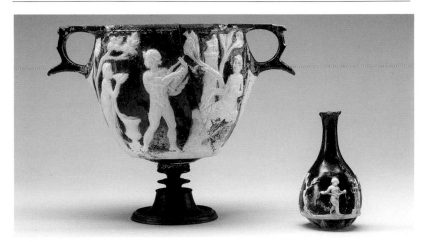

CAMEO GLASS SKYPHOS AND FLASK

Roman, ca. 25 B.C.–A.D. 25
Glass
H: (skyphos) 10.5 cm (4 ¹/₈ in.); (flask) 7.6 cm (3 in.)
84.AF.85; 85.AF.84

Cameo glass vessels, surviving in small numbers, exhibit the great skill of the
artisans responsible for their difficult manufacture and carved decoration. This
two-handled cup, or skyphos, is decorated with Dionysiac imagery: satyrs with
female figures on either side. Its foot is a modern reconstruction. The imagery
on the scent bottle includes Erotes, a striding pharaoh, an obelisk, and an altar
surmounted by the Egyptian god Thoth. These motifs may relate to the story of
Horus, who was brought back to life by Thoth.

CUP WITH REMOVABLE LINER

Asia Minor, Ionia (?), second century B.C.
Silver with gilding
12.1 x 9.1 cm (4¾ x 3⅝ in.)
87.AM.58

Fashioned in the shape of a young bull's head, this cup was made in two parts: an outer case of raised silver decorated with bold repoussé designs and punch work, and a liner that was smoothly polished to hold liquid and fabricated to fit snugly inside the neck of the bull. The animal's eyes were inlaid with glass paste and the ears, separately made. The mouth and tear ducts were covered with leaf gilding as were the immature horns and the fillet around the throat. The gilded horns and fillet identify this young bull as a sacrificial animal. The cup was used for drinking wine or for pouring it as a libation to the gods.

HANDLE OF AN OINOCHOE

Greek, Thrace (?), second century B.C.
Silver with gilding
27 x 9.5 cm (10⁹⁄₁₆ x 3¾ in.)
85.AM.163

This figure of a Triton, or merman, was once the handle of a pouring vessel. His scaly tail was attached at the top by the calyx of gilded acanthus leaves that surrounds his hips and at the base by his flipped-back fins. In his left hand he once held the trident of a sea god.

RHYTON

Parthian, first century B.C.
Gilt silver
24 x diam. (rim) 12.1 cm (9⅛ x 4¾ in.)
86.AM.752.1

The horn of this luxury vessel terminates in the gilded body of a snarling lynx. An Aramaic inscription on the gilded rim tells us the (undeciphered) name of the silversmith responsible for the production of this fantastic vessel as well as the weight of the silver used in its manufacture. Depictions of rhyta like this which survive on earlier painted vases suggest that they were used to aerate wine, which would have been poured through the horn and the body of the lynx, exiting through the spout concealed between his legs. The wine was caught in a drinking cup and consumed.

BOWL

Parthian, first century B.C.
Silver with gilding and garnets
Diam. 20.2 cm (7⅞ in.)
86.AM.752.3

The interior of this bowl is lavishly decorated with a wide variety of elegant gilded floral patterns, which are inset with garnets. The florals are divided in a geometric pattern by silver bands. In the center is an elaborate gilded floral calyx containing a large central garnet. The pentagonal decorative system with florals is characteristic of Greek Hellenistic luxury vessels, but the central calyx is modeled on Near Eastern decorative patterns. This combination of styles suggests that the bowl was produced in an atelier located at the eastern edge of the Hellenistic world.

RHYTON
Parthian, 50 B.C.–A.D. 50
Gilt silver inlaid with glass paste
27.4 x diam. (rim) 12.6 cm (10¾ x 4¹⁵/₁₆ in.)
86.AM.753

This gilt silver rhyton is one of the most elaborately decorated drinking horns to survive from antiquity. Relief floral patterns cover the entire surface of the curving horn, which terminates in the forequarters of an antlered stag. Executed with exceptional attention to anatomical detail—particularly apparent in the veins running down the creature's snout—the stag is shown in flight. The preserved inlaid eyes of glass paste with the whites clearly visible heighten the expression of fear and alertness conveyed by the stag's raised head. Though the vessel surely was produced in a workshop on the eastern fringes of the Hellenistic world, the closest parallels for the ornamental designs are found among the late Hellenistic decorative arts created under the Seleucid rulers who inherited the eastern part of Alexander the Great's sprawling empire.

The term *rhyton* comes from the Greek verb meaning "to run through." Liquid contents poured in at the open rim would have run out in a fine stream through the spout between the stag's forelegs.

A fine dotted inscription in Aramaic preserved on the underside of the stag's body dedicates this rhyton to the goddess Artemis.

PLATE SHOWING AN OLD FISHERMAN

Late Antique
Silver with gilding
Diam. 60 cm (23½ in.)
83.AM.347

This silver plate may have been found together with the one illustrated below. They are splendid examples of the continuation of Hellenistic traditions in late antiquity, although they have considerably different styles and subject matter.

This plate shows an old fisherman removing a fish from his hook. In the background, two fish hang from a wall; the rest of his catch spills out of baskets around him. A line below his feet represents the edge of the sea, beneath which a variety of marine animals swim. The composition is enhanced by gilding and is surrounded by a rim of gold.

PLATE SHOWING A PHILOSOPHICAL DEBATE BETWEEN PTOLEMY AND HERMES

Late Antique
Silver
45 x 28.6 cm (17¾ x 11¼ in.)
83.AM.342

An unidentified man is enthroned at the top of this plate. Below, a bearded man, designated by a Greek inscription as Ptolemaeus, is seated on the left; behind him stands a woman designated as Skepsis. Facing them is a man named as Hermes, but the inscription relating to the woman with him is lost. The subject seems to be allegorical, perhaps of the philosophical dispute between Science and Mythology. If this is the case, Ptolemy, the founder of the Alexandrian school of scientific thought, is seen here in debate with Hermes Trismegistos, the exponent of traditional human wisdom as embodied in myth.

REPOUSSE BAND WITH MYTHOLOGICAL SCENES
South Italian, 540–530 B.C.
Gilt silver
Approx. 8.5 x approx. 37 cm (3⅜ x 14½ in.)
83.AM.343

Although small in scale, this band preserves a wealth of mythology. Episodes of combat are arranged in pairs separated by panels depicting running gorgons. From the left, Zeus battles the snake-legged giant, Typhon; Orestes prepares to kill his mother, Klytaimnestra; Perseus is about to behead the Gorgon Medusa in the company of his patroness, Athena; and Theseus battles the Minotaur.

EARRINGS
Etruscan, Cerveteri, late sixth century B.C.
Gold
Diam. 4.7 cm (1⅝ in.)
83.AM.2.1

Among the finest Etruscan products were objects made for personal adornment. In this pair of earrings the goldsmith combined a variety of techniques, including repoussé, decorative wiring, and granulation.

RING STONE WITH A YOUTH AND DOG

Italic, third–second century B.C.
Gold and cornelian
W: (ring) 1.8 cm (³⁄₄ in.)
85.AN.165

The engraved gemstone depicts a youth feeding a dog. With his body twisted almost frontally and one leg behind the other, the youth leans forward on his staff to drop a morsel to the eager animal. Although this motif appears on scarabs of the fifth century B.C., the modeling of the figures and the mannered composition suggest a date in the Hellenistic period.

INTAGLIO WITH HEAD OF THE DORYPHOROS

Roman, first century B.C.
Gold and dark green agate
Approx. (in setting) 22.4 x 17.3 mm
75.AM.61.1

The exquisitely carved profile head, modeled after a famous late fifth-century-B.C. statue of the Doryphoros, or Spear Bearer, by Polykleitos, exemplifies Roman taste during the late Republic and early Empire for reproductions of the great creations of the classical past.

RING WITH ENGRAVED CAMEO

Roman, late first century B.C.– early first century A.D.
Gold and sardonyx
H: (ring) 2.6 x diam. 2.1 cm
(1 x ¹³⁄₁₆ in.); (stone) 1.9 x 1.1 cm
(³⁄₄ x ⁷⁄₁₆ in.)
87.AN.24

Set within a high-sided gold bezel, this cameo depicts the mythological hero Perseus gazing at the severed head of the Gorgon Medusa. Against his left forearm he cradles his harpa, the short curved sword he used to decapitate the Gorgon. His feet are shod in winged sandals.

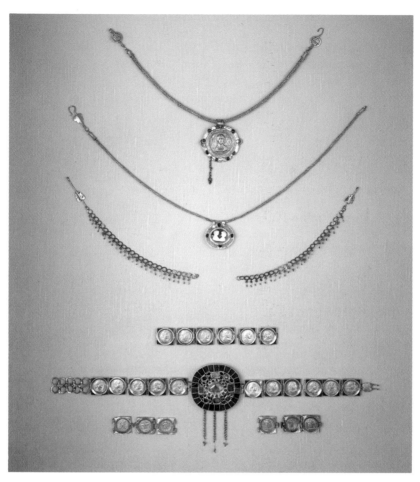

HOARD OF GOLD JEWELRY
Roman, late fourth century A.D.
Gold, precious and semiprecious stones, and glass paste
Various dimensions
83.AM.224–.228

Found together as a hoard, this jewelry presents a fascinating melange of materials and techniques. The individual pieces include two necklaces with large single pendants and two pieces of a third necklace; three openwork bracelets; seven rings; and an elaborate belt. The belt is composed of twenty-two square links made from gold coins of several late fourth-century Roman emperors enframed in green glass; a central medallion decorated with inlays of glass and semiprecious stone and three pendant attachments; and an adjustable hook fastener. The incorporation of preexisting elements into many of the pieces suggests a paucity of artistic creativity at the time of manufacture. At the same time some of the pieces present an interesting aesthetic conflict between the preciousness of the materials employed and the coarseness of their treatment. Despite the artlessness of some elements, the collection clearly evokes the opulence and decadent splendor of the late Roman Empire.

FRAGMENT OF A WALL PAINTING SHOWING A NILOTIC LANDSCAPE

Roman, third quarter of first century A.D.
Tempera on plaster
45.7 x 38.3 cm (18 x 15¼ in.)
72.AG.86

This scene represents the annual flooding of the River Nile. A giant crocodile is about to attack a pygmy on a raft. Battles between pygmies and crocodiles were popular subjects in ancient Greece and Italy for centuries, although whether they were intended as comic or genre scenes has never been fully resolved. The foreground drama is balanced by a monumental colonnaded façade on the far side of the river. Its phantom quality creates the impression that a dense atmosphere fills the entire scene.

PORTRAIT OF A WOMAN

Romano-Egyptian, ca. A.D. 100–125
Encaustic and gilt on wood panel wrapped in linen
55 x 35 cm (21⅞ x 13¾ in.)
81.AP.42

Painting was the most respected form of art in antiquity, but because of the organic nature of the materials from which they were made, the majority of ancient paintings have long since decayed. In the arid desert climate of Egypt, however, a remarkable group of painted mummy portraits has been found. Usually executed from life on a square wooden panel, the portrait was reshaped after the subject's death and incorporated into the linen wrappings of his or her mummy.

This fine example clearly illustrates its funerary use, for parts of the exterior wrappings are preserved. The near-perfect condition of the panel, which has been cleaned but in no way restored, is the result of burial conditions. At the same time, the liveliness, directness, and individuality of the representation support the hypothesis that this lady sat for her portrait sometime before her death. Her hairstyle and jewelry help to date the painting to the early second century A.D.

MANUSCRIPTS

At the J. Paul Getty Museum the art of the Middle Ages is represented primarily by illuminated manuscripts. In 1983 the trustees of the Museum began the collection with the purchase of the holdings of Dr. Peter and Irene Ludwig of Aachen, Germany. The Ludwig manuscripts represent the history of the art of book illumination from the ninth to the sixteenth century. In order to provide as complete and balanced a representation as possible, the Museum is developing this body of manuscripts further. About forty acquisitions have been made since 1984, including codices and cuttings or groups of leaves from individual books.

Illuminated manuscripts were written and decorated entirely by hand. The codex, or bound book, replaced the roll as the most popular vehicle for the written word at the dawn of the Christian era. Over the course of the Middle Ages manuscripts became the most important means of preserving not only scriptures and liturgy but also history, literature, law, philosophy, and science, both ancient and Medieval. Certain books, mainly religious service books in the beginning, came to be decorated with gold, silver, and colors and with elaborate initials, narrative scenes (called miniatures), and other painted decoration.

The Ludwig collection includes masterpieces of Ottonian, Romanesque, Gothic, late Gothic, and Renaissance illumination made in Germany, France, Belgium, Italy, England, Spain, Poland, and the eastern Mediterranean. Among the highlights are three Ottonian manuscripts; the Helmarshausen Gospels, from Romanesque northern Germany (see p. 69); an English Gothic Apocalypse (see p. 73); two Byzantine Gospel books; and a trove of Flemish manuscripts from the fifteenth and sixteenth centuries. Subsequent additions to the Flemish holdings include *The Visions of Tondal,* attributed to Simon Marmion and commissioned by Margaret of York, Duchess of Burgundy (see p. 82), and the *Mira calligraphiae monumenta* that Joris Hoefnagel illuminated for Emperor Rudolf II (see p. 87). The Ludwigs' focus on German and Central European illumination has been enhanced by the purchase of an Ottonian gospel lectionary from Reichenau or Saint Gall (see p. 66), two full-page miniatures from a Gothic psalter from Würzburg, and a copy of Rudolf von Ems' *Weltchronik* from Bavaria (see p. 77). The modest holdings in late Medieval French manuscripts have been amplified by illuminations by the Boucicaut Master (see p. 78), Pseudo-Jacquemart, Jean Fouquet (see p. 83), and Jean Bourdichon, among others.

The manuscripts are exhibited on a rotating basis year-round.

GOSPEL LECTIONARY

German, Reichenau or Saint Gall, third quarter of tenth century
Vellum, 212 leaves (ill.: fol. 4v, decorated initials "IN")
27.7 x 19.1 cm (10^{13}/$_{16}$ x 7^{3}/$_{8}$ in.)
Ms. 16; 85.MD.317

The beginning of each Gospel passage in this manuscript is highlighted by an
elaborate initial entwined with gold and silver vines. Such decorated letters are
characteristic of the monastic scriptoria at Reichenau and Saint Gall, where
many of the finest manuscripts produced in the Ottonian empire—named for
the dynasty founded in 936 by Otto I of Saxony—were made, including books
for the emperors themselves. The initials in this manuscript are unusually delicate examples of this style.

SACRAMENTARY

French, Abbey of Saint Benoît-sur-Loire, first quarter of eleventh century
Vellum, ten leaves (ill.: fol. 9, decorated initial "D")
23.2 x 17.8 cm (9⅛ x 7 in.)
Ms. Ludwig V 1; 83.MF.76

Sacramentaries, the most important type of liturgical book used in the early
Medieval Church, contain the prayers said by a priest at High Mass. The initial
letters of the main prayers in this fragment, such as the "D" shown here, are
filled with elaborate interlace ornament and foliate sprays that hark back to
ninth-century Carolingian models based on Late Antique motifs. The lavish
use of gold, silver, and purple in the decoration, which also includes a full-
page *Crucifixion,* suggests that the book was made for a member of the
royal circle, possibly by Nivardus of Milan, an artist known to have worked
for the French court.

SACRAMENTARY

German, Mainz or Fulda, second
quarter of eleventh century
Vellum, 178 leaves (ill.: fol. 19v,
The Two Marys at the Tomb, and cover)
26.6 x 19.1 cm (10 7/16 x 7 1/2 in.)
Ms. Ludwig V 2; 83.MF.77

The seven full-page miniatures in this
book of prayers to be said at Mass
reveal the monumentality of Ottonian
painting. Set against colored bands,
these scenes from sacred history are
infused with an otherworldly, timeless
quality. The manuscript's luxurious
binding in silver and copper gilt,
showing Christ in Majesty in high
relief, dates from the twelfth century.

GOSPEL BOOK

German, Abbey of Helmarshausen, ca. 1120–1140

Vellum, 168 leaves (ill.: fol. 9v: *Saint Matthew Writing His Gospel*)

22.8 x 16.4 cm (9 x 6⁷⁄₁₆ in.)

Ms. Ludwig II 3; 83.MB.67

The portraits of the evangelists (Matthew, Mark, Luke, and John) in this manuscript exemplify the Romanesque style developed at the abbey of Helmarshausen, an important artistic center in northern Germany that enjoyed the
patronage of Henry the Lion, Duke of Saxony and Bavaria (r. 1139–1180). The
compositions are constructed from lively patterns, and the colors are warm and
saturated. The arrangement of the drapery folds, the facial features, and the use
of firm black outlines to separate areas of different colors have much in common with the work of Roger of Helmarshausen, a famous metalsmith active
at the abbey in the early twelfth century. Saint Matthew is shown here writing
his Gospel.

BREVIARY
Written by Sigenulfus
Italian, Abbey of Montecassino, ca. 1153
Vellum, 328 leaves (ill.: fol. 138v, decorated initial "c")
19.1 x 13.2 cm (7 9/16 x 5 3/16 in.)
Ms. Ludwig IX 1; 83.ML.97

Sigenulfus, the scribe of this breviary, identified himself in a prayer he penned on one of its leaves. He was a member of the monastic community at Montecassino, the principal center of the Benedictine order, founded in the sixth century by Saint Benedict. During the eleventh and twelfth centuries the monastery's scriptorium developed a distinctive style of decorative initial in which the paneled framework of the letter brims over with a frenzied mixture of interlaced tendrils and fantastic creatures.

NEW TESTAMENT

Written by Theoktistos
Byzantine, Constantinople, 1133
Vellum, 280 leaves (ill.: fol. 106v, *Saint John the Evangelist*)
22 x 18 cm (8⅝ x 7¼ in.)
Ms. Ludwig II 4; 83.MB.68

The portraits of the evangelists in this New Testament are among the finest achievements of Byzantine illumination of the Comnenian period (named for the emperors who ruled from 1018 until 1185). The artist's use of line and surface pattern to create expressive effects is typical of the pictorial style of the mid-twelfth century and is also to be found in monumental art.

PSALTER

German, Würzburg, ca. 1240–1250

Vellum, 192 leaves (ill.: fol. 111 v, *Pentecost*)

22.6 x 15.7 cm (8 ⅞ x 6 ³/₁₆ in.)

Ms. Ludwig VIII 2; 83.MK.93

Richly illuminated psalters enjoyed widespread popularity in the thirteenth century as a form of deluxe private devotional book. The abundant decoration of this psalter includes twelve illuminated calendar pages, each decorated with the full-length figure of a prophet; six full-page miniatures depicting imagery from the Old and New Testaments; and ten elaborate, large historiated initials.

The finely chiseled features of the figures and the angular folds of their draperies are typical of thirteenth-century German illumination as practiced in Thuringia and Saxony, in the north, and in Würzburg, in the southwest, which was influenced by the northern schools. The forms of this so-called Zigzag style are remarkably expressive.

APOCALYPSE WITH COMMENTARY BY BERENGAUDUS
English, London, ca. 1250
Vellum, forty-one leaves (ill.: fol. 6, *The Opening of the First Seal: The First Horseman*)
31.9 x 22.5 cm (12½ x 8⅞ in.)
Ms. Ludwig III 1; 83.MC.72

A fashion for richly illustrated copies of the Apocalypse, Saint John's vision of the end of the world as recorded in the New Testament Book of Revelation, developed in England in the mid-thirteenth century. In manuscripts such as this one, which contains eighty-two half-page miniatures, aristocratic lay patrons could contemplate the calamitous events Saint John described through pictures as well as text.

Remarkable for their lively interpretation of the saint's vision, the miniatures show him witnessing the extraordinary events as they occur. The figures were outlined in pen and ink and then modeled with thin, colored washes. Their elegant poses, the graceful contours of the forms, and the decorative patterns of the drapery folds typify early English Gothic illumination at its finest.

VIDAL DE CANELLAS
Feudal Customs of Aragon (Vidal Mayor)
Transcribed and illuminated by Michael Lupi of Çandiu
Spanish, Pamplona, first quarter of fourteenth century
Vellum, 277 leaves (ill.: fol. 72v, *King Jaime I of Aragon Oversees a Court of Law*)
36.5 x 24 cm (14⅜ x 9⁷⁄₁₆ in.)
Ms. Ludwig XIV 6; 83.MQ.165

This sumptuous volume is the only known copy of the law code of Aragon
which King Jaime I (r. 1214–1276) ordered Vidal de Canellas, the bishop of
Huesca, to compile in 1247. The manuscript, a translation into the vernacular
(Navarro-Aragonese), probably was made for one of Jaime's royal successors.
It is illustrated with ten large and numerous small scenes set into initials. The
style of illumination, painted in primary colors, emulates Parisian court art
in both the settings and the slender figural and facial types.

ANTIPHONAL

Italian, Bologna, end of thirteenth century
Vellum, 243 leaves (ill.: fol. 2, *Christ in Majesty*)
58.2 x 40.2 cm (22$^{15}/_{16}$ x 15$^{7}/_{8}$ in.)
Ms. Ludwig VI 6; 83.MH.89

This manuscript is the first volume of a multivolume antiphonal, a compilation
of chants for the Divine Office, the services celebrated at regular intervals
throughout the day. Like all liturgical choir books, its format is large so that all
members of the choir could see the music it contains. It may have been made for
the monastery of San Jacopo di Ripoli, outside Florence, since the same artist
provided that religious community with at least two other choir books. He
probably trained in Bologna, a major center of Gothic art in Italy. The facial
types, abstract patterns of drapery highlights, and certain iconographic motifs
in this antiphonal reflect the impact of Byzantine art. The brilliant luminosity
and monumental figure style achieved by the artist link his work with the
most important Bolognese manuscripts produced at this time.

LIFE OF SAINT HEDWIG OF SILESIA (Hedwig Codex)
Silesian, Court Atelier of Duke Ludwig I of Liegnitz and Brieg, 1353
Vellum, 204 leaves (ill.: fol. 12v, *Saint Hedwig with Donors*)
33.8 x 24.5 cm (13⁵/₁₆ x 9⁵/₈ in.)
Ms. Ludwig XI 7; 83.MN.126

The earliest illustrated *Life of Saint Hedwig*, whose subject was a thirteenth-century duchess of Silesia (a duchy on the border between Poland and Germany), this manuscript contains one of the masterpieces of Central European manuscript illumination: a full-page miniature of Saint Hedwig herself. Hedwig, who founded a number of religious houses in Silesia, carries a book, a rosary, and a small statue of the Madonna, all references to her devout character. The small-scale donor portraits represent Duke Ludwig I of Liegnitz and Brieg and his wife, Agnes, the saint's descendants who commissioned the manuscript in 1353.

RUDOLF VON EMS

Weltchronik
Bavarian, ca. 1405–1410
Vellum, 309 leaves (ill.: fol. 89v, *Moses' Vision of the Back of God's Head;
Moses' Shining Face and the Ark of the Covenant*)
33.5 x 23.5 cm (13⅛ x 9¼ in.)
Ms. 33; 88.MP.70

The *Weltchronik* of Rudolf von Ems is a Medieval chronicle written in rhymed couplets that weaves biblical, classical, and other secular texts into a continuous history of the world beginning with Creation. Of the numerous surviving copies of the text, the Museum's copy, with 364 miniatures, is the most lavishly illustrated. This miniature depicts two scenes: an unusual representation of Moses' vision of the back of God's head (Exodus 33.12–23) and Moses after his descent from Mount Sinai (Exodus 34.29–35). He is represented with horns, the result of a Medieval misreading of the original Hebrew text.

The vigorous, broadly painted miniatures with contrasting, often incandescent colors find few parallels in European art of the first half of the fifteenth century. The values of the International Gothic Style that dominated European art at this time favored more finely proportioned figures, elegant, flowing drapery, and softer colors (see p. 79). The miniatures of the *Weltchronik* have a bold narrative style; the figures have intense expressions and interact dramatically. The identity of the original owner, undoubtedly an important nobleman, is unknown. Duke Albert IV of Bavaria (r. 1463–1508) and Duchess Kunigunde owned the manuscript later in the fifteenth century.

BOOK OF HOURS
Illuminated by the Boucicaut Master and his atelier
French, Paris, ca. 1415–1420
Vellum, 281 leaves (ill.: fol. 254, *All Saints*)
20.4 x 14.9 cm (8 x 5⅞ in.)
Ms. 22; 86.ML.571

Paris was one of the great artistic centers of Europe at the beginning of the
fifteenth century. The Boucicaut Master, the city's leading illuminator, is named
for the book of hours he illuminated for Jean le Meingre (called Boucicaut), the
marshal of France (Paris, Musée Jacquemart-André). In this miniature accompanying a prayer to all saints, the individualized physiognomies of the crowded
assembly of saints, their alert expressions, and the sense of subtle movement
breathe life into a stock theme of late Medieval art. The elegant, flowing draperies and lush color epitomize the courtly taste of this era.

CRUCIFIXION
Attributed to the Master of Saint Veronica
German, Cologne, ca. 1400–1410
Vellum
23.6 x 12.5 cm (9 5/16 x 4 15/16 in.)
Ms. Ludwig Folia 2; 83.MS.49

The courtly elegance of the International Gothic Style is captured in the rich costumes, brilliantly diapered backgrounds, and softly modeled forms of this *Crucifixion* and its companion miniature, *Saint Anthony*. Attributed to the Master of Saint Veronica, a leading artist of Cologne, these leaves may once have been part of a liturgical manuscript.

MISSAL

Illuminated by the Master of the Brussels Initials

Italian, Bologna, ca. 1389–1404

Vellum, 259 leaves (ill.: fol. 172a, *The Calling of Saint Andrew*)

33 x 24 cm (13 x 9⁷/₁₆ in.)

Ms. 34; 88.MG.71

The Master of the Brussels Initials, an Italian artist who is credited with the sumptuous illumination in this missal—a service book used to celebrate Mass—was a key figure in the development of the International Gothic Style. The border style seen here, with its exuberant foliage inhabited by whimsical creatures, was introduced by him into Paris and was adopted by many French artists in the early fifteenth century. The missal was commissioned by the Bolognese cardinal Cosmato Gentili di Meliorati and acquired subsequently by the anti-pope John XXIII, whose coat of arms appears in the lower margin.

BOOK OF HOURS
Illuminated by the Master of Guillebert de Mets and others
Flemish (Tournai ?), ca. 1450–1460
Vellum, 286 leaves (ill.: fol. 18v, *Saint George and the Dragon*)
19.4 x 14 cm (7¾ x 5½ in.)
Ms. 2; 84.ML.67

This deluxe manuscript, with gigantic, brightly colored foliage wending
through its borders, exemplifies the inventive, often playful character of the
decorated borders in Flemish manuscripts. The acanthus leaves dwarf Saint
George and the princess whom he rescues from the dragon. The anonymous
illuminator was a leading Flemish artist in the first half of the fifteenth century.

ample ouuerte et moult   nece qui essoient illeas
obscure / Ceste vallee estoit   bruslece et arsee et fuict

THE VISIONS OF TONDAL

Illuminations attributed to Simon Marmion; written by David Aubert
Flemish, Valenciennes and Ghent, 1474
Vellum, 45 leaves (ill.: fol. 13v, *The Valley of the Homicides*)
36.3 x 26.2 cm (14⁵/₁₆ x 10⁵/₁₆ in.)
Ms. 30; 87.MN.141

The Visions of Tondal is a Medieval cautionary tale about a wealthy Irish knight
whose soul embarks on a dreamlike journey through hell, purgatory, and
heaven, during which he learns the wages of sin and the value of penitence.
Written in Latin in the middle of the twelfth century, the popular story was
translated into fifteen European languages over the next three centuries. This
sumptuous manuscript, commissioned by Margaret of York, Duchess of Burgundy, is the only known illuminated copy of the text and contains twenty
miniatures. In the scene illustrated here, Tondal's guardian angel shows the soul
of the knight a burning valley in which murderers' souls are cooked on an
iron lid. Above it the slim, naked figures of the damned rise and fall in a graceful ballet of torment.

HOURS OF SIMON DE VARIE

Illuminated by Jean Fouquet, the chief associate of the Bedford
Master, and the Master of Jean Rolin II
French, Paris or Tours, 1455
Vellum, 97 leaves (ill.: fol. 2v, *Heraldic Illustration: Arms of Simon de Varie*)
11.5 x 8.2 cm (4⁹/₁₆ x 3¹/₄ in.)
Ms. 7; 85.ML.27

This heraldic miniature and the three other frontispiece miniatures in this book
of hours are the work of Jean Fouquet, one of the great fifteenth-century
French painters. Influenced by Italian Renaissance painting, the figural style
and compositions in Fouquet's miniatures have a geometric simplicity and
monumentality that transformed French art. His technical mastery is evident in
the flesh tones of the woman, rendered in points of pure color invisible to the
naked eye, and in the transparent cascade of hair through which the background
of the miniature can be seen. Fouquet, a panel painter as well as an illuminator,
worked successively at the courts of Charles VII (r. 1422–1461) and Louis XI
(r. 1461–1483) of France.

This manuscript is a portion of a book of hours made for Simon de Varie, a
high official at the royal court. De Varie's mottos appear in the borders of the
miniatures. In this one "Vie à mon Desir," an anagram of his name, is inscribed
on the collar of the heraldic dog supporting the arms. Separated from the Getty
volume in the seventeenth century, the two other portions of the book are now
in the Royal Library at the Hague.

GUALENGHI-D'ESTE HOURS

Illuminations attributed to Taddeo Crivelli and Guglielmo Giraldi
Italian, Ferrara, ca. 1470
Vellum, 211 leaves (ill.: fol. 3v, Taddeo Crivelli, *The Annunciation*)
10.8 x 7.9 cm (4¼ x 3⅛ in.)
Ms. Ludwig IX 13; 83.ML.109

This tiny, richly illuminated book of hours is one of the finest examples of
manuscript illumination from the school of Ferrara, whose ruling family, the
Este, were ambitious bibliophiles and patrons of the arts. The humanist values
of their court and, more generally, the Renaissance culture that flourished in
much of Italy are reflected in the classical details of the architecture, carved
stone inscriptions, and putti in the borders. Like other painters from Ferrara,
Crivelli used deep shadows and bright highlights to model figures, giving them
their characteristic sculptural quality. The manuscript probably was made for
Orsina d'Este, who married Andrea Gualengo in 1469.

SPINOLA HOURS

Illuminated by Gerard Horenbout and others
Flemish, Ghent or Malines, ca. 1515
Vellum, 312 leaves (ill.: fol. 92v, *The Annunciation*)
23.2 x 16.6 cm (9⅛ x 6½ in.)
Ms. Ludwig IX 18; 83.ML.114

In one of the most inventive page designs of sixteenth-century manuscript illumination, Gerard Horenbout has unified the traditionally separate zones of miniature and decorated border. The main subject, the Annunciation, shown taking place in the Virgin Mary's chamber, is portrayed in the middle of the page. The exterior of the building, where angels joyously pick flowers in the Virgin's garden, is depicted in the surrounding border. Thus the setting of the miniature has expanded beyond its traditional boundaries to fill the page and incorporate additional devotional stories and events.

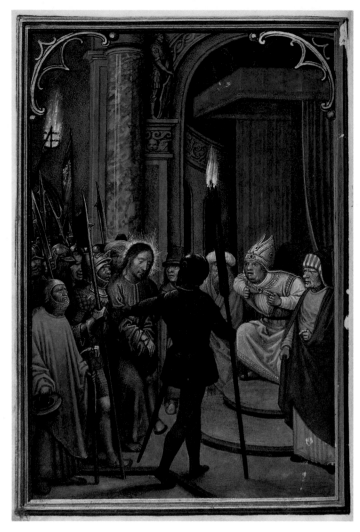

PRAYER BOOK OF CARDINAL ALBRECHT OF BRANDENBURG

Illuminated by Simon Bening
Flemish, Bruges, ca. 1525–1530
Vellum, 337 leaves (ill.: fol. 128v, *Christ before Caiaphas*)
16.8 x 11.5 cm (6⅝ x 4½ in.)
Ms. Ludwig IX 19; 83.ML.115

The Prayer Book of Albrecht of Brandenburg represents one of the pinnacles in the career of the illustrious Flemish illuminator Simon Bening. The forty-one full-page miniatures in this book dramatically recount Christ's childhood and ministry followed by his persecution, Crucifixion, and Resurrection. The artist inspires the reader's compassion and understanding through the depiction of Christ as a vulnerable man who suffered continuous emotional and physical abuse at the hands of his tormentors.

MIRA CALLIGRAPHIAE MONUMENTA (Model Book of
Calligraphy)
Written by Georg Bocskay (Hungarian); illuminated by Joris Hoefnagel
(Flemish), Vienna, 1561–1562; 1591–1596
Vellum, 150 leaves (ill.: fol. 74, *Carnation and Walnut*)
16.6 x 12.4 cm (6⁹/₁₆ x 4⁷/₈ in.)
Ms. 20; 86.MV.527

Originally written for the Holy Roman Emperor Ferdinand I (r. 1556–1564),
this volume is a compendium of display scripts by one of the great scribes of the
Renaissance. The calligraphy on each page represents a different style of writing
inventively arranged. Thirty years later, Emperor Rudolph II (r. 1576–1612),
Ferdinand I's grandson, commissioned Hoefnagel to illuminate the same pages.

Jan Van
Huysum fecit 1722

PAINTINGS

During his lifetime J. Paul Getty purchased paintings from every major European school of art between the thirteenth and twentieth centuries. He did not, however, consider himself a collector of paintings. His writings, especially his diaries, repeatedly name the decorative arts as his first love. He also felt, not without some justification, that the best paintings were already in museums. Significant pieces of furniture, however, could still be acquired, and for much less money than a first-class painting.

Nevertheless Mr. Getty began his painting and decorative arts collections at about the same time: during the 1930's. By World War II he owned two major pictures, Gainsborough's *Portrait of James Christie* and Rembrandt's *Portrait of Marten Looten* (see p. 5), as well as a number of lesser paintings. Mr. Getty again began to collect paintings actively in the 1950's, although it was not until the middle of the following decade that he attempted to buy pictures that could be said to match the stature of individual pieces in his decorative arts holdings. Curiously, he never bought paintings to complement the French furniture he collected with such enthusiasm.

During the last decade of his life and until his death in 1976, Mr. Getty gave funds to the Museum that allowed it to acquire paintings from periods he had not previously found interesting. Some of these pictures are of considerable importance. When Mr. Getty left the bulk of his estate to the Museum, the opportunity presented itself to acquire major works on a wider scale. In the past eight years these pictures, which number nearly half of the paintings on exhibit, have significantly transformed the galleries.

During Mr. Getty's years as a collector, he succeeded in gathering a representative group of Italian Renaissance and Baroque paintings, plus a few Dutch figurative works of importance. In recent years these collections have been expanded. The French and Dutch schools have been much strengthened, and later French pictures by the Impressionists and their successors have been added. As a result, there is a greater overall balance of holdings from the earlier periods until the end of the nineteenth century. More significantly, the works that can now be displayed are often more representative and important examples than in the past.

It is expected that the Museum will continue to acquire and exhibit paintings of the highest quality during the coming years, to complete the collection's transformation from a fairly small personal holding into a moderately sized but exceptionally choice public repository.

PACINO DI BONAGUIDA
Italian, active until the 1340's
The Chiarito Tabernacle, 1340's
Gilded gesso and tempera on panel
101.4 x 113.3 cm (39⅞ x 44⅝ in.)
85.PB.311

This Florentine triptych is unusual for its combination of media and its depiction of the visionary experiences of a layman, Chiarito del Voglia, for whom it was created. The symbolic representation of the Communion of the Apostles was executed in gilded relief to contrast with the painted narrative scenes. At the bottom of the central panel, the patron's visions while attending Mass indicate that he shares in the mystical communion above.

SIMONE MARTINI
Italian, ca. 1284–1344
Saint Luke, after 1333
Tempera on panel
56.5 x 37 cm (22¼ x 14½ in.)
82.PB.72

Saint Luke, author of one of the Gospels, painted a portrait of the Virgin and Christ. Painters adopted him as their patron saint and named their guilds and academies after him. Here Luke is depicted as the author of his Gospel, with his emblematic beast, the winged ox. This panel, in its original jeweled frame, flanked a central panel of the Virgin and Child. Its rich materials and the curvilinear composition reflect Byzantine influence, but the precision of line and refined characterization are Martini's contribution to later Sienese painting.

GENTILE DA FABRIANO
Italian, ca. 1370–1427
Coronation of the Virgin, 1420's
Tempera on panel
87.5 x 64 cm (34½ x 25½ in.)
77.PB.92

Probably painted for a church in Fabriano, Italy, Gentile's *Coronation* once formed part of a double-sided standard carried on a pole during religious processions. Stylistically it occupies the borderline between the essentially Gothic International Style and the early Florentine Renaissance. Traces of the former linger in the richly patterned surfaces of the hangings and robes, which, however, clothe solidly constructed figures. The folds create a sense of volume as well as pattern, and the figures gesture with convincing naturalism.

MASACCIO (Tommaso Guidi)
Italian, 1401–1428
Saint Andrew, 1426
Tempera on panel
52.3 x 33.2 cm (21 x 12 in.)
79.PB.61

This image of Saint Andrew holding his traditional symbols, a cross and book, comes from the only securely documented altarpiece by Masaccio, a polyptych painted in 1426 for the private chapel of Giuliano da San Giusto, a wealthy notary, in the church of the Carmine in Pisa. The panel belonged on the right side of the upper tier, so that Andrew once appeared to gaze sadly toward the central panel depicting the Crucifixion.

The *Saint Andrew* exemplifies Masaccio's contribution to the founding of an early Renaissance painting style. The single figure is endowed with volume and solidity; his simplified draperies fall in sculptural folds. Since the panel stood high above the ground, the figure is slightly foreshortened for viewing from below. Furthermore, Masaccio has abandoned the elegant detachment of earlier styles in favor of a naturalistic treatment of emotion. It was this powerful combination of three-dimensional form and bold expression which so impressed his contemporaries.

VITTORE CARPACCIO
Italian, 1455/56–1525/26
Hunting on the Lagoon, ca. 1490
Oil on panel
75.4 x 63.8 cm (30 x 29 in.)
79.PB.72

On one side this painting shows bow-
men hunting birds; on the other, one
finds a trompe l'oeil painting of a letter rack. The marks of hinges on the
latter side indicate that this panel once
functioned as a decorative shutter
or cabinet door. It is a fragment of
a composition that included two
women on a balcony with the lagoon
visible between them.

BARTOLOMMEO VIVARINI
Italian, ca. 1432–1499
Polyptych with Madonna and Child, Saint James Major, and Various Saints, 1490
Tempera on panel
280 x 215 cm (110¼ x 84⅝ in.)
71.PB.30

Shown in this altarpiece are Saints Catherine and Ursula, the Virgin and Child,
and Saints Apollonia, Mary Magdalen, John the Baptist, John the Evangelist,
James Major, Bartholomew, and Peter.

ANDREA MANTEGNA
Italian, ca. 1431–1506
Adoration of the Magi, ca. 1495–1505
Distemper on linen
54.6 x 70.7 cm (21½ x 27⅞ in.)
85.PA.417

Paintings of the Adoration of the Magi traditionally provided the opportunity to create grandiose scenes of Oriental richness. But for this private devotional work, Mantegna exhibited the spirit of Renaissance humanism by concentrating on the psychological interactions among the principal characters in the sacred narrative. The close-up scheme of half-length figures compressed within a shallow space and set before a neutral background was created by the artist from his study of ancient Roman reliefs, reflecting the Renaissance interest in reviving classical antiquity.

The magi, or kings, represent the three parts of the then known world—Europe, Asia, and Africa—and personify royal recognition of Christ's superior kingship. They offer their gifts in vessels of carved agate, Turkish tombac ware, and an early example of Chinese porcelain in Europe. The subject was particularly appropriate for the painting's probable patrons, the Gonzaga family, rulers of Mantua who employed Mantegna as court painter from 1459 until his death.

For this intimate scene Mantegna chose a technique unusual for his time. Rather than oil or egg tempera, the artist employed thin washes of pigment suspended in animal glue. The result is an image with dazzling color and a matte surface emphasizing the refined draughtsmanship and brushwork. Paintings in this technique usually were left unvarnished, as is the case with the *Annunciation* by Bouts (see p. 103). However, Mantegna's painting was varnished at a later date. Now mostly removed, the varnish darkened the image somewhat, although the refinement of painting and the richness of detail still attest to Mantegna's mastery.

DOSSO DOSSI (Giovanni de' Luteri)
Italian, active 1512–1542
Mythological Scene, early 1520's
Oil on canvas
160 x 132 cm (63 x 52 in.)
83.PA.15

As court painter at Ferrara in northern Italy, Dosso was famous for his enchanting and mysterious landscapes inhabited by gods and legendary mortals. His dreamlike, ornamental compositions embody the idealized Renaissance image of classical antiquity developed earlier by the Venetian painter Giorgione.

Just as his work on this mythological scene neared completion, Dosso changed his mind about its subject and painted a landscape over the draped female figure now visible at the center left. The three remaining figures—the sleeping nude, the protective central figure, and the god Pan on the right—may have represented the story of Pan and the nymph Echo. During the nineteenth century the fourth figure, whose significance remains enigmatic, was detected under the landscape and revealed by scraping away the upper paint layer.

PONTORMO (Jacopo Carucci)
Italian, 1494–1557
Portrait of Cosimo I de' Medici, ca. 1537
Oil or oil and tempera on panel transferred to canvas
92 x 72 cm (36¼ x 28⅜ in.)
89.PA.49

This swaggering image of the eighteen-year-old Cosimo I was painted shortly after he won a battle for the leadership of the Florentine state. Cosimo became one of the most important patrons of the sixteenth century, and this portrait is the first example of his use of painting for purposes of propaganda.

Pontormo's refined, complex play of colors and oval forms is prototypical of Mannerist portraiture, as are the sitter's aristocratic poise and elegance. The image is also an unusually subtle psychological study. Its appeal centers on the facial expression and the artist's ability to render both the arrogance and vulnerability of youth.

FRANCESCO SALVIATI
(Francesco de' Rossi)
Italian, 1510–1563
Portrait of a Bearded Man, ca.
1550–1555
Oil on panel
109 x 86.3 cm (43 x 34 in.)
86.PB.476

The bold pose, affected gesture, strident colors, and glossy finish of this likeness are characteristic of mid-sixteenth-century Mannerism. Such frozen images of aristocratic rank developed from portraits like Pontormo's of Cosimo de' Medici (see p. 96). However, this likeness reflects the influence of Pontormo's greatest pupil, Bronzino, whose portraits of the Medici court Salviati studied in the 1540's.

GIOVANNI GEROLAMO SAVOLDO
Italian, ca. 1480–after 1548
Shepherd with a Flute, ca. 1525
Oil on canvas
97 x 78 cm (38³/₁₆ x 30¹¹/₁₆ in.)
85.PA.162

Paintings that idealize the life of rustic folk were popularized by Venetian masters in the early 1500's just as pastoral poetry and drama also began to flower. Created for an urban audience, these works eulogize the presumed simplicity of rural life, often questioning the permanence of city institutions. Savoldo's shepherd indicates that his thatched-roof farmhouse is built against the ruins of an earlier civilization.

VERONESE (Paolo Caliari)
Italian, 1528–1588
Portrait of a Man, ca. 1560
Oil on canvas
193 x 134.5 cm (76 x 53 in.)
71.PA.17

Based on the approximate date of the
canvas, the age of the subject, and the
physical resemblance to known
portraits of Veronese, this painting
has sometimes been identified as a
self-portrait. However, full-length
portraits of artists were uncommon in
the sixteenth century, and the privi-
lege of wearing a sword was rarely
granted to painters. Whatever his
identity, it has recently been suggested that the sitter may have had
some connection with the church of
San Marco in Venice, which appears
at the lower left.

DOMENICHINO (Domenico Zampieri)
Italian, 1581–1641
Christ Carrying the Cross, ca. 1610
Oil on copper
53.7 x 68.3 cm (21⅛ x 26⅝ in.)
83.PC.373

Domenichino continually sought ideal form and grandeur, known as *disegno*,
in his compositions; this painting is one of the earliest realizations of his goal.
He often painted on copper plates, using small, controlled brush strokes to
describe the massive volumes of the figures. As they lean over the fallen Christ,
the tormentors embody the inexorable forces driving them to Calvary. Intent
on their cruelty, they do not acknowledge Christ's pleading silence.

GIOVANNI LANFRANCO
Italian, 1582–1647
Moses and the Messengers from Canaan, 1624
Oil on canvas
218 x 246.3 cm (85¾ x 97 in.)
69.PA.4

Painted in Rome about fourteen years after Domenichino's much smaller *Christ
Carrying the Cross*, Lanfranco's composition displays the same concern for monumentality. His stark, powerful work shows the spies returning to Moses (at the
left with his characteristic staff) laden with grapes and other proofs of the
fruitfulness of Canaan (Numbers 13:21–25). This picture formed part of a set
of nine paintings (the Museum owns one other) representing Old and New
Testament miracles involving food and hence prefiguring the Eucharist.

GUERCINO (Giovanni Francesco
Barbieri)
Italian, 1591–1666
Portrait of Pope Gregory XV,
ca. 1622–1623
Oil on canvas
133.4 x 97.8 cm (52½ x 38½ in.)
87.PA.38

Baroque portraits of exalted sitters
often concentrate on the trappings of
power without aiming to represent
character. This picture is a remarkable
exception. Instead of merely portray-
ing the pontiff in his official capacity,
Guercino depicted him sympathetically
in the last months of his life, worn by
the cares of office and failing health.

BERNARDO CAVALLINO
Italian, 1616–ca. 1656
The Shade of Samuel Invoked by Saul, 1650's
Oil on copper
61 x 86.5 cm (24 x 34 in.)
83.PC.365

Cavallino's late works on copper display the strange, brilliant coloring and delicate brushwork suited to fantastical subjects. In this scene the shrouded ghost of
Samuel at the left warns the kneeling king that he will be killed if he goes to war
against David and the Philistine army the next morning (I Samuel 28:7–25).
The witch of Endor, who has summoned the ghost, studies the king's face for
his reaction. This painting may have been inspired by the unstable political
situation in Naples.

PIER FRANCESCO MOLA
Italian, 1612–1666
Vision of Saint Bruno, 1660–1666
Oil on canvas
194 x 137 cm (76 3/8 x 53 7/8 in.)
89.PA.4

Saint Bruno was the founder of the Carthusian order, a monastic community committed to solitary meditation as the most effective means of attaining union with God. Here the saint demonstrates this hermetic ideal, reaching to touch a vision that has appeared during his devotions in the wilderness.

   Mola won fame in Rome for his rich landscapes with dramatic cloud formations based on Venetian examples. In this work he also has demonstrated his exceptional ability to convey complex psychological states as he explores one of the most important themes of Baroque art: humanity's confrontation with the divine.

CANALETTO (Giovanni Antonio Canal)
Italian, 1697–1768
View of the Dogana, Venice, 1744
Oil on canvas
60.3 x 96 cm (23³/₄ x 37³/₄ in.)
83.PA.13

The Dogana, or customs warehouse of Venice, is situated at the point of land where the Grand Canal and Giudecca Canal meet. Behind the Dogana is the church of Santa Maria della Salute, whose massive dome is seen here rising in the background at the right. The waterways are crowded with gondolas and other vessels. The ship at the center flies an English flag, perhaps in deference to one of Canaletto's many English patrons. The artist's renderings of famous Venetian landmarks made his work irresistible to travelers passing through Italy on the Grand Tour; they purchased his paintings as mementoes of their voyages. Canaletto's paintings are especially prized for their masterful rendering of atmosphere and dramatic compositional effects.

DIERIC BOUTS
Flemish, active ca. 1445–d. 1475
The Annunciation, ca. 1450–1455
Distemper on linen
90 x 74.5 cm (35⁷⁄₁₆ x 29³⁄₈ in.)
85.PA.24

This painting represents the Incarnation, or conception of Christ, thought to
have occurred when the archangel Gabriel announced to the Virgin Mary,
"You shall conceive and bear a son, and you shall give him the name Jesus"
(Luke 1:31). Mary, seated on the ground to indicate her humility, is interrupted
at her prayers and responds with a gesture conveying her acknowledgment
of the miracle and acceptance of God's will. Bouts created a composition of surprising simplicity, minimizing symbolic detail to establish a mood of hushed
solemnity. This image was executed in an equally austere technique. Thin
washes of pigment suspended in animal glue were applied to fine linen to give
the scene a delicate, luminous quality. This was the first scene in an altarpiece
relating the life of Christ; the final scene, the *Resurrection*, is in the Norton
Simon Museum, Pasadena.

MASTER OF THE PARLEMENT DE PARIS
Flemish or French, second half of fifteenth century
Crucifixion, mid–fifteenth century
Oil on panel
48 x 71.5 cm (18⅞ x 28¼ in.)
79.PB.177

This panel and its two wings depicting the arrest of Christ and the Resurrection narrate the events of Christ's Passion and rising from the dead. Christ appears three times on this panel. On the left Saint Veronica wipes his face with her handkerchief as he carries the cross to Calvary; he is crucified in the center; and on the right he releases souls from Purgatory.

JAN GOSSAERT
(called MABUSE)
Flemish, ca. 1478–1532
Portrait of Francisco de los Cobos y Molina (?), ca. 1530–1532
Oil on panel
43.8 x 33.7 cm (17¼ x 13¼ in.)
88.PB.43

The most accomplished Flemish painter of his generation, Gossaert combined keen observation of form, facile rendering of surface textures, and psychological insight to create portraits remarkable for their strong physical presence. This one probably represents Francisco de los Cobos, a major Spanish patron who was the powerful secretary and chief financial adviser to Emperor Charles V (r. 1519–1556).

ANTHONY VAN DYCK
Flemish, 1599–1641
Portrait of Agostino Pallavicini, ca. 1625–1627
Oil on canvas
216 x 141 cm (85⅛ x 55¼ in.)
68.PA.2

During his second visit to Genoa in 1625–1627, Van Dyck is known to have painted Agostino Pallavicini dressed in the robes of an ambassador to the pope. The sketchy curtain behind the sitter here bears the Pallavicini coat of arms; his dramatic robes surely were intended for ceremonial duties. Van Dyck emulated his master Rubens' rich and dramatic painting style but brought to portraiture a unique aristocratic refinement that transformed the genre.

PETER PAUL RUBENS
Flemish, 1577–1640
The Meeting of King Ferdinand of Hungary and the Cardinal-Infante Ferdinand of Spain at Nördlingen, 1635
Oil on panel
49.1 x 63.8 cm (19 5/16 x 25 1/8 in.)
87.PB.15

In 1635 the Cardinal-Infante Ferdinand, brother of King Philip IV of Spain (r. 1621–1665) and newly appointed governor of the southern Netherlands, made his triumphal entry into Antwerp. In celebration, the city's magistrates commissioned a series of temporary structures to line the parade route from Peter Paul Rubens.

This sketch served as a model for Rubens' shop in the production of a large canvas that decorated the Stage of Welcome. It depicts the meeting of the Cardinal-Infante and his Habsburg cousin shortly before their combined armies scored an important victory over Protestant forces in 1634. Since the Spanish Habsburgs recently had lost the northern Netherlands to the Protestant cause, this subject promoted the role of dynastic alliance in the preservation of the Catholic faith.

In a typically Baroque manner, the historical personages are accompanied by allegorical figures who comment on the significance of the event. The meeting took place on the Danube, represented here as a classical river god sitting on an urn flowing with water and blood, an allusion to the coming battle. To the right Germania gazes mournfully at the viewer as a winged genius draws her attention to the meeting and the promise of peace.

Sketches like this have been highly valued since Rubens' day; executed entirely by the artist, they are direct records of a great master creating a complex, dramatic composition in a spirited, economical way.

JOACHIM WTEWAEL
Dutch, 1566–1638
Mars and Venus Surprised by the Gods, ca. 1606–1610
Oil on copper
20.3 x 15.5 cm (8 x 6⅛ in.)
83.PC.274

In this small painting Wtewael has depicted a scene from Ovid's *Metamorphoses* (IV.171ff.) with appropriate raucous humor. As Vulcan (at the bottom right) draws his forged net from the bed, the lovers Mars and Venus (Vulcan's wife) reel back, Cupid and Apollo raise the canopy for a peek, and a gleeful Mercury (wearing a winged cap) looks up to Diana in the clouds at the right. Saturn (holding a sickle) and Jupiter crane their necks to behold the embarrassed adulterers. The rhythmically paired heroic nude figures show Wtewael at the height of his inventive powers as a Mannerist artist.

This tiny cabinet picture may have been painted for the private enjoyment of a connoisseur familiar with Ovid's text and capable of appreciating both the artist's skill and the significance of the painting's symbolic details.

HENDRICK TER BRUGGHEN
Dutch, 1588–1629
Bacchante with a Monkey, 1627
Oil on canvas
102.9 x 90.1 cm (40½ x 35½ in.)
84.PA.5

This devotee of Bacchus, with her fruit, nuts, and monkey (a symbol of appetite or gluttony), may represent the sense of taste. Caravaggio had set the fashion for half-length figures representing the senses in Rome at the turn of the century, and his works certainly influenced Ter Brugghen's both in type and style. Coming from a Catholic city in largely Protestant Holland, Ter Brugghen was one of the few Dutch artists to visit Italy in an era of religious controversy and so became instrumental in the spread of new Italian styles to the North. The lively, unidealized portrayal of the model and the sculptural modeling of the flesh in darks and lights are characteristic of the caravaggesque style Ter Brugghen adopted while in Rome (1604–1614) and carried back to his native Utrecht.

PIETER JANSZ. SAENREDAM

Dutch, 1597–1665

The Interior of Saint Bavo, Haarlem, 1628

Oil on panel

38.5 x 47.5 cm (15¼ x 18¾ in.)

85.PB.225

Pieter Saenredam, the great renovator of architecture painting in the Netherlands, transformed the idealized, artificial perspective of earlier pictures done mostly by Flemish artists. By minimizing anecdote and adopting an unusually low viewpoint, Saenredam concentrated on the depiction of light, color, and space in monumental buildings.

The Interior of Saint Bavo, Haarlem is the earliest known painting by Saenredam. It is the first of twelve paintings of the interior of this church—one of the finest Gothic buildings in Holland—which he painted between 1628 and 1660. Saenredam conceived his views of church interiors with the help of detailed, on-the-spot studies, extensive measurements, and perspective drawings. He introduced a high degree of realism, as if inviting the viewer to walk into the buildings he depicted, by choosing a low, somewhat off-center, eye-level viewpoint. In this painting, however, the artist transcended mere architectural portraiture by merging two distinct views of the church: from the north transept straight ahead into the south transept, and to the east, into the choir. Different shades of white paint—ranging from bluish-white to creamy yellow—suggest the northern and southern light shining into the building. Remarkably, Saenredam substituted an altarpiece for the doors in the church's south transept and included a stained glass window with a depiction of the Immaculate Conception. These elements of Catholic worship, which would not have been found in a whitewashed Dutch Protestant church, suggest a Catholic commission for this painting.

REMBRANDT VAN RIJN
Dutch, 1606–1669
An Old Man in Military Costume,
ca. 1630
Oil on panel
66 x 50.8 cm (26 x 20 in.)
78.PB.246

From the beginning of his career,
Rembrandt was occupied with historical subjects. He soon took up a
related genre, which he made his own:
studies of individual costumed figures.

Such works as *An Old Man in
Military Costume* are not primarily
portraits but rather studies in which
Rembrandt explored the human
condition.

PHILIPS KONINCK
Dutch, 1619–1688
Panoramic Landscape, 1665
Oil on canvas
138 x 167 cm (54½ x 65½ in.)
85.PA.32

Koninck's expansive view, like Rembrandt's drawings and prints of landscapes
(see p. 145), combines real and imagined elements to create the illusion of
sweeping space as if seen from a great height.

REMBRANDT VAN RIJN
Dutch, 1606–1669
Saint Bartholomew, 1661
Oil on canvas
86.5 x 75.5 cm (34⅛ x 29¾ in.)
71.PA.15

Rembrandt's subject has been identified as Saint Bartholomew by virtue of the knife, instrument of his martyrdom, gripped in his right hand, but the actual sitter may have been one of the artist's friends or neighbors in Amsterdam. Rembrandt painted several deeply moving studies of such men in the guise of saints and apostles in the early 1660's. Somberly colored and expressively painted, the *Saint Bartholomew* is an examination of age and introspection. The momentary distraction of the *Old Man in Military Costume* (see p. 110) has given way to the saint's profound absorption; his thoughts are far removed from the present. During the thirty years between the *Old Man* and this painting, Rembrandt's style also shifted from concern for surface effects to a probing analysis of form and structure.

JACOB VAN RUISDAEL
Dutch, 1628/29–1649
The Sluice, 1648–1649
Oil on panel
39.4 x 55.9 cm (15½ x 22 in.)
86.PB.597

Seen from a low viewpoint and faintly illuminated by sunbeams, a humble
stone bridge and a sluice gain unexpected monumentality in this painting.

AELBERT CUYP
Dutch, 1620–1691
A View of the Maas at Dordrecht, ca. 1645–1646
Oil on panel
50 x 107.3 cm (19¾ x 42¼ in.)
83.PB.272

The contemplative calm of this picture is enhanced by the artist's subdued palette of browns, whites, and greens. Thin layers of color applied in sweeping
horizontal brush strokes describe the limpid surface of the river, while the
thickly painted white clouds evoke the dense atmosphere of Holland.

JAN STEEN
Dutch, 1626–1679
The Drawing Lesson,
ca. 1665
Oil on panel
49.3 x 41 cm (19 3/8 x 16 1/4 in.)
83.PB.388

FRANS
VAN MIERIS
THE ELDER
Dutch, 1635–1681
Pictura, 1661
Oil on copper
12.5 x 8.5 cm
(5 x 3 1/2 in.)
82.PC.136

These two compositions are both allegories of the painter's art. The Steen
shows the interior of a painter's studio, cluttered with symbolic objects, while
the painter, his pupil, and his apprentice concentrate on correcting a drawing.
Several of the symbolic objects in *The Drawing Lesson* also appear as attributes
of *Pictura,* a personification of the art of painting. Along with her palette,
brushes, and statue, this figure wears a mask hung on a chain around her neck.
The mask may be an emblem of deceit or illusion (the painter creates an illusion
of reality), or it may refer to the dramatic arts (like the dramatist, the painter
must invent and "stage" convincing scenes).

PAULUS POTTER
Dutch, 1625–1654
The Piebald Horse,
ca. 1650–1654
Oil on canvas
49.5 x 45 cm
(19½ x 17¾ in.)
88.PA.87

GERARD TER BORCH
Dutch, 1617–1681
Interior of a Stable,
ca. 1654
Oil on panel
45.3 x 53.5 cm
(17¹³⁄₁₆ x 21¹⁄₁₆ in.)
86.PB.631

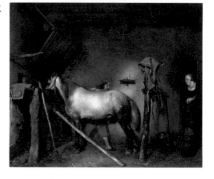

A horse was valuable property in seventeenth-century Holland. Both of these paintings reflect the pride the prosperous Dutch took in their costly livestock. Potter, the most successful animal painter of his time, painted his magnificent piebald horse as a monumental creature, attesting to its nobility. The horseman in the background, returning with his groom and dogs to his country house, may be the animal's proud owner. Ter Borch's *Interior of a Stable* is roughly contemporary. A fine riding horse is represented in a barn, tended by a man dressed in distinctly middle-class attire like the woman at the right. This subject was highly unusual for Ter Borch, normally a painter of sophisticated genre scenes and portraits.

JAN VAN
HUYSUM
Dutch, 1682–1749
Vase of Flowers, 1722
Oil on panel
80.5 x 61 cm
(31 ¾ x 24 in.)
82.PB.70

AMBROSIUS
BOSSCHAERT
THE ELDER
Dutch, 1573–1621
Flower Still Life, 1614
Oil on copper
28.6 x 38.1 cm
(11 ¼ x 15 in.)
83.PC.386

The mania for cultivating flowers in Baroque Europe fostered the development and popularity of flower painters. One of the earliest of these specialists, Ambrosius Bosschaert the Elder, has grouped flowers from different seasons—roses, tulips, forget-me-nots, cyclamen, a violet, and a hyacinth—along with insects equally beautiful and short-lived. Jan van Huysum's bouquet of a century later includes several of the same flowers and insects as well as a bird's nest. Both painters celebrated the decorative qualities of their subjects but also intended to provoke serious reflection on the transience of life and beauty.

GEORGES DE LA TOUR
French, 1593–1652
Beggars' Brawl, ca. 1625–1630
Oil on canvas
94.4 x 142 cm (37 ¼ x 56 in.)
72.PA.28

Although the subject of this early La Tour composition—a scuffle between two beggar musicians—is clear, its meaning continues to elude us. Scholars have suggested that the two troupes are fighting for possession of a lucrative street corner; that the scene was taken from the theater, possibly from a comedy; that the musician in velvet is unmasking the false blindness of the hurdy–gurdy player by squirting lemon juice in his eye; and that the painting is really an illustration of a moralizing proverb such as "Wretched is he who can find no one more wretched than himself." La Tour has illuminated his frieze of beggars with a cold light from overhead. His strange palette, his attention to unusual textures such as the tanned and wrinkled skin of the aged beggar, and his extraordinarily detached treatment of his characters made La Tour one of the most original and enigmatic artists of the seventeenth century.

VALENTIN DE BOULOGNE
French, 1591–1632
Christ and the Adulteress, 1620's
Oil on canvas
168 x 220 cm (66 x 86½ in.)
83.PA.259

Although Valentin seems to have drawn his characters from seventeenth-century taverns and guardrooms, they participate here in a precise depiction of an episode from the New Testament (John 8:3–8): "And the scribes and Pharisees brought unto him a woman...and...said...Master, this woman was taken in adultery, in the very act. Now Moses in the law commanded us, that such should be stoned: but what sayest thou?...Jesus stooped down, and with his finger wrote on the ground, as though he heard them not. So when they continued asking him, he...said unto them, He that is without sin among you, let him first cast a stone at her...."

PHILIPPE DE CHAMPAIGNE
French, 1602–1674
Portrait of Antoine Singlin, ca. 1646
Oil on canvas
79 x 65 cm (31⅛ x 25⅝ in.)
87.PA.3

Around 1643 Philippe de Champaigne became involved in the Jansenist religious movement. Antoine Singlin was a Jansenist priest admired for his forthright sermons and the simple rigor of his piety. The virtues Singlin embodied are reflected in the austerity and directness of this portrait.

NICOLAS POUSSIN
French, 1594–1665
The Holy Family, ca. 1651
Oil on canvas
100 x 132 cm (39¾ x 53 in.)
81.PA.43 (Owned Jointly with the Norton Simon Museum, Pasadena)

The traditional Holy Family, which includes the Virgin and Child with Saint
Joseph, has here been extended to include the youthful Saint John the Baptist
with his mother, Saint Elizabeth. Framed by the passive figures of their parents,
the holy children reach out in a lively embrace. Six putti bearing a basket of
flowers, a ewer, a towel, and a basin of water may prefigure Saint John's later
baptism of Christ. Painted in Poussin's late classicizing style, *The Holy Family*
reflects his study of Raphael and antique sculpture. Poussin's intellectual
approach to painting, his insistence on harmony among all parts of the composition, and his concentration on the ideal and the abstract contrast with the
dramatic style of the caravaggist painter Valentin (see p. 117).

JEAN-SIMEON CHARDIN
French, 1699–1779
Still Life, ca. 1759–1760
Oil on canvas
37.8 x 46.7 cm (14⅞ x 18⅜ in.)
86.PA.544

For this small still life, Chardin arranged elements of varying color and
texture—a silver goblet, peaches, walnuts, and grapes—on a rough stone ledge
and then set about reproducing the three-dimensional objects on the canvas in a
convincing manner. The result is a vivid depiction of ordinary household items
arranged in a simple, balanced composition. Chardin emphasized the play of
light on the objects, so that the goblet, for example, reflects the rich red of the
peaches and tan of the nuts while also gleaming with reflected light.

CHARLES LE BRUN
French, 1619–1690
The Martyrdom of Saint Andrew,
ca. 1647
Oil on canvas
98.5 x 80 cm (38¾ x 31½ in.)
84.PA.669

Charles Le Brun was the first artist to hold the influential title of first painter to the king under Louis XIV (r. 1643–1715). Le Brun's classical manner owed a great deal to Poussin (see p. 118), whom he had followed to Rome in 1642. This oil study dates from soon after Le Brun's return to Paris and illustrates his admiration for Poussin and the Baroque as well as a knowledge of classical antiquity. In the painting Saint Andrew appeals dramatically to the heavens as he is being tied to the cross of his martyrdom amid a turbulent group of soldiers and mourners. The painting most likely served as the preparatory *modello* for an altarpiece in Notre Dame, Paris. The altarpiece, which differs from this oil study in several details, was one of the yearly gifts to the cathedral known as the "*grand Mai*" and a much coveted commission. Interestingly, the present painting, not the altarpiece, served as the model for a later engraving of the composition.

JEAN-FRANÇOIS DE TROY
French, 1679–1752
Preparations for the Ball, 1735
Oil on canvas
81.8 x 65 cm (32⅜ x 25⁹/₁₆ in.)
84.PA.668

De Troy, a painter of historical, religious, and mythological subjects, is best remembered for his invention of the *tableau de mode* showing elegant company in fashionable interiors. Inspired by the *fêtes galantes* of Watteau and his circle and by Dutch seventeenth-century genre paintings, de Troy depicted a group of gallants preparing for a masked ball. Their lavish dress and the fine furniture add to the opulence of the setting. The mood of hushed expectancy is enhanced by the flickering candlelight which illuminates the room.

This canvas and its pendant, *Return from the Ball* (location unknown), were commissioned in 1735 for Germain-Louis de Chauvelin, minister of foreign affairs and keeper of the seal under Louis XV (r. 1715–1774).

JEAN-BAPTISTE GREUZE
French, 1725–1805
The Laundress, 1761
Oil on canvas
40.6 x 31.7 cm (16 x 12⅞ in.)
83.PA.387

"This little laundress is charming, but she's a rascal I wouldn't trust an inch," wrote the critic Denis Diderot when he saw this painting exhibited at the Salon of 1761. The instant rapport Greuze's painted characters achieve with their flesh-and-blood audience was the source of his success with the exhibition-going public of his time and remains one of his great strengths today. The theatrical and moralizing qualities of his work are of primary importance: where will this young servant's bold appraisal of the viewer lead? Yet the painting's technical qualities also deserve notice. Just as sensual as the laundress' glance and exposed ankle are the creamy brush strokes delineating the folds of laundry, warm ceramic and wood surfaces, gleaming copper pots, and the woman's youthful skin.

JEAN-ETIENNE LIOTARD
Swiss, 1702–1789
Portrait of Maria Frederike van Reede-Athlone, 1755–1756
Pastel on vellum
53.5 x 43 cm (21 x 17 in.)
83.PC.273

Liotard endeavored to record three-dimensional objects as faithfully as possible, a technique he called "deception painting." His definition of volume through subtle gradations of color and his brilliant description of surface textures create the startling realism that sets his work apart from that of his contemporaries. Here he has captured the alert intelligence of a seven-year-old girl, daughter of an aristocratic Dutch family, with amazing immediacy. By setting his subject against a plain background, Liotard focused attention on the details of the sitter, her clothing, and her pet dog. He also suggested an enigmatic shyness on the part of the girl, whose eyes are averted, implying a complex personality about to bloom. In contrast to the refined elegance of the child is the eager curiosity of the dog, which stares unabashedly at the viewer. Utilizing the difficult medium of pastel, Liotard masterfully evoked both the physicality and psychology of his subject.

THOMAS GAINSBOROUGH
English, 1727–1788
Portrait of James Christie, 1778
Oil on canvas
126 x 102 cm (49⅝ x 40⅛ in.)
70.PA.16

James Christie, founder of the London auction house that still bears his name, is depicted as if attending one of his own sales. Much praised at the time of its exhibition at the Royal Academy, the portrait epitomizes Rococo elegance and grace. The style's characteristic arabesques are found in Christie's relaxed pose, in the tree and foliage of the Gainsborough landscape he leans against, in the gilt frame of the painting behind it, and in the artist's accomplished, free handling of paint throughout the composition.

FRANCISCO JOSE DE GOYA Y LUCIENTES
Spanish, 1746–1828
Portrait of the Marquesa de Santiago, 1804
Oil on canvas
209 x 126.5 cm (82½ x 49¾ in.)
83.PA.12

An English visitor to Spain described the Marquesa de Santiago as "very profligate and loose in her manners and conversation, and scarcely admitted into female society.... She is immensely rich." Goya has not attempted to flatter his sitter in this portrait; its beauty depends on the manner in which it is painted. Broad brush strokes create the deep and splendid landscape; the dress is matte black on black. Thick impasto simulates the glitter of gold braid on the sleeve, and the diaphanous lace mantilla is described with broad, flat strokes of shimmering color. Goya's expressionistic painting style sets him apart from most other artists of the age, who sought ideal beauty in their subjects and high finish in their technique.

THEODORE GERICAULT
French, 1791–1824
Portrait of a Man, ca. 1818–1819
Oil on canvas
46.5 x 38 cm (18¼ x 15 in.)
85.PA.407

This portrait may represent Joseph, the professional artist's model who posed for Gericault's masterpiece *The Raft of the Medusa* (Paris, Musée du Louvre), completed in 1819. A haunting character study never used in the larger work, this portrait belongs to an important group of paintings and drawings of blacks by Gericault.

THEODORE GERICAULT
French, 1791–1824
The Race of the Riderless Horses, 1817
Oil on paper laid on canvas
19.9 x 29.1 cm (7¾ x 11½ in.)
85.PC.406

Gericault devoted much of his year in Rome (1816–1817) to planning a monumental painting of the race of the riderless horses, the climax of the Roman Carnival season. The furious struggle between lunging animals and their handlers at the start of the race may symbolize the Romantic artist's search for a balance between his own passionate feelings and need for professional discipline.

JACQUES-LOUIS DAVID
French, 1748–1825
The Farewell of Telemachus and Eucharis, 1818
Oil on canvas
87.2 x 103 cm (34½ x 40½ in.)
87.PA.27

David found the story of the romance between Telemachus, son of the homeric hero Odysseus, and the nymph Eucharis in the didactic novel *Télémaque*, written by the French prelate Fénelon in 1699. The idealized lovers, physically perfect and morally pure, take refuge in a grotto for a last embrace before Telemachus departs in search of his errant father. *The Farewell*'s erotic theme, sumptuous coloring, and sensually painted half-length figures are characteristic of David's late history paintings executed in Brussels after his exile from France in 1816. His saturated reds and blues and vividly convincing flesh tones—reflecting the influence of the Flemish Baroque artist Rubens—are here married to the Greek perfection of line and form characteristic of late Neoclassicism.

PIERRE-PAUL PRUD'HON
French, 1758–1823
Justice and Divine Vengeance Pursuing Crime, ca. 1805/06
Oil on canvas
32 x 41 cm (12⅞ x 16⅛ in.)
84.PA.717

Prud'hon painted this dramatic study in preparation for a large canvas destined to hang in the criminal courtroom of the Palace of Justice in Paris. Its subject, apparently inspired by a line from Horace—"Retribution rarely fails to pursue the evil man" (*Odes* III.2.31–32)—reminded the courtroom's occupants of the inevitability of justice. Divine Vengeance, hunting by torchlight, prepares to deliver a fleeing robber to her inexorable companion Justice, armed with sword and scales.

   By the time he painted this rapid yet assured oil sketch, Prud'hon had settled on the major elements of his composition in a series of drawings. Odd elements such as the disproportionately large size of the naked victim seem to have been chosen deliberately to balance the design. In the colored oil, Prud'hon experimented with the play of moonlight and torchlight across the figures as well as with the pose of the victim, which he changed slightly in the final version. The thin, almost transparent layers of paint impart a ghostly reality to the scene. After completing the sketch, the artist studied individual figures in drawings and then executed the monumental canvas (Paris, Musée du Louvre). Not surprisingly, the sense of rushing motion and nocturnal mystery evoked by the spontaneous sketch was lost in the carefully finished final version.

JEAN-FRANÇOIS MILLET
French, 1814–1875
Man with a Hoe, 1860–1862
Oil on canvas
80 x 99 cm (31½ x 39 in.)
85.PA.114

When Millet first exhibited *Man with a Hoe* at the Salon of 1863, it raised a storm of controversy which lasted well into the twentieth century. The artist, fatalistic in his outlook, intended his image of an exhausted peasant to represent the dignity and strange beauty of manual labor. By contrasting the bent, awkward figure with the richly painted broken earth of the field, Millet was calling attention to the costs and benefits of agrarian life. The division of the foreground into distinct patches of stony, thorny terrain on the one hand and cultivated earth on the other graphically portrays the results of the peasant's toil. However, the starkness of the scene and harsh characterization of the peasant were attacked by most of the critics who reviewed the 1863 exhibition. Moreover, since it concluded a long series of similar subjects Millet painted during a decade of heated public debate over the fate of the peasant in French society (and of slavery in America), *Man with a Hoe* was taken by some to represent the socialist position on the laboring poor. In response to claims that his painting was both ugly and politically radical, Millet responded, "Some tell me that I deny the charms of the country. I find more than charms, I find infinite glories.... But I see as well, in the plain, the steaming horses at work, and in a rocky place a man, all worn out, whose 'han!' has been heard since morning, and who tries to straighten himself for a moment and breathe. The drama is surrounded by beauty."

PIERRE-AUGUSTE RENOIR
French, 1841–1919
La Promenade, 1870
Oil on canvas
81.3 x 65 cm (32 x 25½ in.)
89.PA.41

Renoir's promenading couple heads for the undergrowth. Although the subject owes its inspiration to Watteau, Courbet, and Monet, the technique of using different brushwork for different purposes was entirely Renoir's. Thus he defined the white dress with long, flowing strokes but created the landscape with short touches that skitter and dance on the surface. Light, atmosphere, and the diagonal movement of the lovers unify the composition. The artist's vigorous enjoyment of pleasure—both amorous and painterly—animates every aspect of this early Impressionist work.

PAUL CEZANNE
French, 1839–1906
Portrait of Anthony Valabrègue,
ca. 1869–1871
Oil on canvas
60 x 50 cm (23⅝ x 19¾ in.)
85.PA.45

Cézanne, the poet and critic Anthony Valabrègue, and the writer Emile Zola grew up together in Aix-en-Provence. Their friendship persisted into adulthood, and each became a subject for the others' work, although they did not always find the results flattering. Valabrègue remarked that Cézanne, in painting his friends, seemed to be avenging himself for some hidden injury, and Zola's treatment of Cézanne in his novel *L'Oeuvre* (1876) damaged their friendship.

Cézanne depicted Valabrègue three times between 1866 and ca. 1875. This portrait, the second in the series, endows the slim poet with an imposingly solid appearance and an expression of introspective concern. Thickly applied slabs of paint, especially in the sitter's face and beard, add to the sense of an almost sculptural presence, as does the severely restricted palette. The resulting work of art, while not the accurate description of appearance one would expect from a more conventional painter, is an absorbing, powerful evocation of mood.

EDOUARD MANET
French, 1832–1883
The Rue Mosnier with Flags, 1878
Oil on canvas
65.5 x 81 cm (25 ¾ x 31 ¾ in.)
89.PA.71

On June 30, 1878, the day of a national celebration, Manet painted two canvases of his street with its decoration of flags. This deceptively casual rendering, executed as late afternoon shadows advanced down the rue Mosnier, captures the scene with brilliant economy. Virtuoso brushwork describes the scene with minimal fuss and maximum effect. Manet's commitment to subjects from modern life makes this painting more than an urban landscape, however; the poignant juxtaposition of the *quartier's* one-legged beggar with elegant façades and carriages along the street suggests the inequities of life in the new urban environment.

EDGAR DEGAS
French, 1834–1917
Waiting, ca. 1882
Pastel on paper
48.2 x 61 cm (19 x 24 in.)
83.GG.219 (Owned Jointly with the Norton Simon Museum, Pasadena)

Unlike his contemporaries the Impressionists, Degas arranged his subjects into compositions of great formal beauty laden with emotional power. He returned frequently to favorite poses or figures, finding fresh significance in each new combination, viewpoint, or variation. In this pastel Degas may have intended the juxtaposed figures on the bench to contrast the brilliant, artificial world of the theater with the drabness of everyday life. If the suggestion is correct that the scene shows a dancer and her mother from the provinces awaiting an audition at the Paris Opera, then a contrast between the ephemeral glamor of the opera dancer and the dreary respectability of the provincial may also be implied. Each woman is absorbed in her own thoughts, yet they share a sense of tension and anticipation.

JAMES ENSOR
Belgian, 1860–1949
Christ's Entry into Brussels in 1889, 1888
Oil on canvas
252.5 x 430.5 cm (99½ x 169½ in.)
87.PA.96

James Ensor's enormous canvas detailing Christ's entry into the capital city of his native Belgium presents us with a complex and terrifying vision of the society of the future. Christ, whose face appears to be the artist's self-portrait, rides a small donkey in the midst of a teeming procession of Mardi Gras revelers. Many in the crowd wear masks, Ensor's device for suggesting the deceptiveness and artificiality of modern life.

The canvas' bold color contrasts and thick crusts of paint give the impression of spontaneity, yet Ensor planned every figure in advance. In both subject and technique this masterpiece shocked nineteenth-century viewers, and it could not be exhibited until 1929. Yet its reputation brought many visitors to the artist's studio, and its visual and emotional impact contributed to the development of Expressionism early in the twentieth century.

EDVARD MUNCH
Norwegian, 1863–1944
Starry Night, 1893
Oil on canvas
135.2 x 140 cm (53⅜ x 55⅛ in.)
84.PA.681

The coastline of Åsgårdstrand has changed little since Munch painted this view looking toward the Oslo fjord. The white rail fence still encloses an orchard of fruit trees, and the massed foliage of three tightly clustered lindens still looms above them. The summer night sky filled with stars and reflections—and with the reddish planet Venus hovering on the horizon—held a special symbolism for Munch, who associated it with memories of an unhappy youthful love affair. Thus, despite its surprising accuracy as a landscape, this painting actually portrays a mood: romantic melancholy tinged with mysticism. Munch used the composition to introduce a series of images called the Frieze of Life, which treats the subjects of love and death in mythic, autobiographical terms.

VINCENT VAN GOGH
Dutch, 1853–1890
Irises, 1889
Oil on canvas
71 x 93 cm (28 x 36⅝ in.)
90.PA.20

Van Gogh painted *Irises* in the garden of the asylum at Saint-Rémy, where he was recuperating from a severe attack of mental illness. Although he considered the work more a study than a finished picture, it was exhibited at the Salon des Indépendants in September 1889. Its energy and theme—the regenerative powers of the earth— express the artist's deeply held belief in the divinity of art and nature. However, the painting's vivid color contrasts, powerful brushwork, and friezelike composition reflect Van Gogh's study of other artists, notably Paul Gauguin and the Japanese master Hokusai.

DRAWINGS

The Museum's drawings collection was begun in July 1981, after Mr. Getty's death, with the purchase of Rembrandt's red chalk study of a nude woman as Cleopatra (see p. 145). However, it was not until a year later that the Museum started to expand beyond this initial acquisition, and the Department of Drawings was formed. The collection now contains approximately 350 drawings ranging from the second half of the fifteenth through the end of the nineteenth century. The major objective is to build a representative historical collection, with special emphasis on the most important and brilliant draughtsmen in the Western tradition.

The Museum decided to develop a drawings collection because drawing is perhaps the most universal of all art forms. The genre embraces a wide variety of media, including pen and ink, different chalks, charcoal, watercolor, and so on. Drawing is probably the only artistic area in which virtually everyone has had some personal experience. Moreover, every painter, sculptor, architect, and printmaker has used drawings in studying nature or the work of other artists, in developing ideas, and as preparatory devices in evolving projects. In addition, some drawings have been made intentionally as completed works of art.

The drawings illustrated in this handbook demonstrate great diversity. The two Dürers in the collection consist of a highly finished and self-contained watercolor and gouache study of a stag beetle (see p. 132) and, by contrast, a more loosely drawn pen study of the Good Thief for a painting or print of the Crucifixion (see p. 133). There are a number of compositional studies for paintings, such as the works illustrated by Veronese, Poussin, and David; studies of individual figures, such as the Carracci, Watteau, and Rubens drawings; preparatory studies for stained glass windows by Altdorfer and Baldung; a preparatory study for a print by Piranesi; and a number of drawings made for their own sake as completed works of art, such as those by Bernini, Rembrandt (the landscape), Blake, Ingres, Daumier, Van Gogh, and Cézanne. It is not always possible to know the reason why an artist made a drawing; this is true of the Leonardo and the Rembrandt nude. However, irrespective of intention, drawings bring one closer to the variety of an artist's inner thoughts and instincts than any other art form.

Drawings are sensitive to light and therefore cannot be exhibited on a continuous basis. Those on exhibition at any given time represent selections from the collection united by period, origin, or other common elements.

ALBRECHT DÜRER
German, 1471–1528
Stag Beetle, 1505
Watercolor and gouache
14.2 x 11.4 cm (5 ⁹/₁₆ x 4 ½ in.)
83.GC.214

This startling image is characteristic of the artist's interest in nature and coincides with the more definitively scientific renderings of plant and animal life by Dürer's Italian contemporary Leonardo da Vinci. At the same time, this drawing presents a living form illusionistically, with the stag beetle casting a shadow on the plain ground of the paper as if actually crawling across it. The beetle was drawn with exceptional care, especially in regard to the modulations of tone along the creature's back.

ALBRECHT DÜRER
German, 1471–1528
Study of the Good Thief, ca. 1505
Pen and brown ink
26.8 x 12.6 cm (10⁹/₁₆ x 5 in.)
83.GA.360

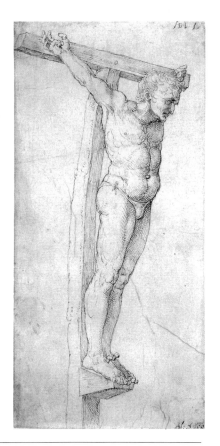

In this drawing Dürer conceived the Good Thief with ingenuity and power. The figure is shown sharply foreshortened and at a rather stark angle, thus heightening the dramatic impact of the form. In addition Dürer has shown the cross bending under the weight of the figure, whose hand, curled in agony, also reinforces the physical and emotive force of the drawing. The use of the pen here was completely assured and developed from the broad statement of the basic image to the clear definition and modeling of the forms.

HANS BALDUNG GRIEN
German, ca. 1484/85–1545
A Monk Preaching, ca. 1505
Pen and brown ink
30.8 x 22.3 cm (12 ⅛ x 8 ¹³/₁₆ in.)
83.GA.194

Baldung was one of the most distinguished of Dürer's followers. His characteristically rhythmic and sharply accented pen strokes were derived from his master. This drawing seems to have been made as a preparatory study for a stained glass window in a church, a type of commission Baldung is known to have undertaken. The preacher is depicted before his congregation. The presence of Christ above his left hand indicates that he is speaking on the subject of salvation or the Last Judgment.

HANS HOLBEIN THE YOUNGER
Swiss, 1497–1543
Portrait of a Cleric or Scholar, ca. 1532–1543
Point of brush and black ink and black and red chalk on pink prepared paper
21.9 x 18.5 cm (8⅝ x 7¼ in.)
84.GG.93

This unidentified man is taken to be a scholar or cleric on account of his
hat. Holbein probably made the portrait during his second English period
(1532–1543). At this time he served as painter to King Henry VIII, and the sitter
may have been connected with the Tudor court. His portliness and pronounced
features lend him considerable presence. The drawing's focal point is the sen-
sitively drawn contour of the face, which Holbein sketched in black chalk and
went over in black ink lines. The face was softly modeled in red chalk, using the
pink of the paper as a middle ground. Outside the face the handling is broad
and painterly, as in the hair and robe. Holbein was attuned to the uniqueness of
the features of each of his sitters, which he rendered with unparalleled precision
and objectivity. For this reason his portrait drawings present sixteenth-century
personalities with greater freshness and fullness than almost any other artist
of his time.

ALBRECHT ALTDORFER
German, ca. 1480–1538
Christ Carrying the Cross, ca. 1510–1515
Pen and black ink, gray wash, and black chalk
Diam. 30.4 cm (11¹⁵⁄₁₆ in.)
86.GG.465

Albrecht Altdorfer was the leading painter of the Danube School, which was
noted for its focus on dynamic motifs such as mountains and forests and on
religious imagery of great dramatic power. This drawing shows Christ fallen
under the weight of the cross and the blows of his tormentors. Utilizing a
highly unusual approach, Altdorfer has shown Christ from behind, a vantage
point that creates a strong spatial thrust. The figures possess the monumentality
one expects in large-scale paintings, as can be seen particularly in the soldier at
the left brandishing a whip. Bravura pen work heightens the emotional resonance of the scene. Altdorfer made this drawing as a design for a stained glass
window; it is the only such drawing by him to have survived.

LEONARDO DA VINCI
Italian, 1452–1519
Three Sketches of a Child with a Lamb, ca. 1503–1506
Pen and brown ink and black chalk
20.3 x 13.8 cm (8 x 5 7/16 in.)
86.GG.725

These sketches—three in ink and three in faint chalk—represent Saint John
the Baptist or, more likely, the Christ child wrestling with a lamb. The multiple studies attest to Leonardo's lifelong fascination with depicting highly
animated subjects. He probably made this drawing in preparation for a painting
of the Virgin and Child with Saint John, now lost but known through copies.
Leonardo inscribed the drawing at the top and on the reverse in his characteristic mirror script.

RAPHAEL (Raffaello Sanzio)
Italian, 1483–1520
Christ in Glory, ca. 1520
Black and white chalk and gray wash on pale gray prepared paper
22.5 x 17.9 cm (8⅞ x 7⅟₁₆ in.)
82.GG.139

Raphael was one of the greatest, and probably the single most influential,
draughtsmen in the history of Western art. This study of Christ in Glory was
made from a live model, and one can still see the underdrawing of the undraped
legs that was rapidly sketched in first. The pose of the figure exemplifies
Raphael's classicism; the grand, idealized form is described with economy and
monumentality. At the same time, the artist used very subtle gradations of light
and shadow to give physical and spiritual resonance to the image.

LORENZO LOTTO
Italian, ca. 1480–1556
Saint Martin Dividing His Cloak with a Beggar, ca. 1530
Brush and gray-brown ink, white and cream gouache heightening, and black
chalk on brown paper
31.4 x 21.7 cm (12⅜ x 8⁹/₁₆ in.)
83.GG.262

Due to the strong perspective angle indicating that one was intended to view
this scene from below, it has been suggested that the drawing was made as a
study for an organ shutter or a fresco above eye level. The dynamic movement
and complex organization anticipate Baroque aesthetics, while the varied media
are typical of sixteenth-century Venetian art.

PAOLO VERONESE
Italian, 1528–1588
Sheet of Studies for the Martyrdom of Saint George, 1566
Pen and brown ink and brown wash
28.9 x 21.7 cm (11 3/8 x 8 9/16 in.)
83.GA.258

This drawing is made up of a series of freely sketched studies for several sections of Veronese's altarpiece in the church of San Giorgio, Verona, showing the martyrdom of Saint George. This sheet reveals the creative energy of the artist as he explored varying solutions to different figural problems throughout the composition. He employed a combination of pen and wash with great sensitivity, and, although the sheet is made up of diverse parts, it comes together with a sense of the overall rhythmic flow of the artist's hand moving across the page.

ANNIBALE CARRACCI
Italian, 1560–1609
Study of Triton Blowing a Conch Shell,
ca. 1600
Black and white chalk on blue paper
40.7 x 24.1 cm (16 x 9½ in.)
84.GB.48

Annibale made this drawing for his
brother Agostino to use as the basis
for a figure in a fresco by the latter on
the ceiling of the Palazzo Farnese,
Rome. The powerful sculptural form
and the monumentality of the mus-
cular image derive from ancient
Roman sculpture and from Annibale's
study of Renaissance art, while the use
of black and white chalk on blue
paper reflects his interest in sixteenth-
century Venetian art.

GIAN LORENZO BERNINI
Italian, 1598–1680
Portrait of a Young Man, ca. 1625–1630
Red and white chalk
33.2 x 21.8 cm (13¹/₁₆ x 8⅝ in.)
82.GB.137

This dynamic portrait of an unidentified young man is among the finest examples of Bernini's work in this genre. It was drawn with great liveliness, and the
red chalk strokes show remarkable breadth and freedom. The use of white
chalk, applied with equal vitality, adds luminosity and greater substance to the
image. In this portrait Bernini has achieved an active interrelationship between
viewer and subject; the young man looks out with boldness and animation.

L. Carache.   938

GIOVANNI BATTISTA PIAZZETTA
Italian, 1683–1754
A Boy Holding a Pear, ca. 1737
Black and white chalk on blue-gray paper
39.2 x 30.9 cm (15⁷/₁₆ x 12³/₁₆ in.)
86.GB.677

In this drawing Piazzetta has depicted a well-dressed adolescent wearing
a brocade vest, full-sleeved shirt, and feathered cap. The sitter, who holds
a pear, gazes at the viewer provocatively. The same boy, recognizable by his
sweet, lively countenance, appears in a number of Piazzetta's works and has
been identified as his son, Giacomo. This drawing was made as an independent
work of art, to be appreciated for the charm of its subject and sumptuousness of
its technique. Piazzetta applied the chalk in a variety of densities, as can be seen
in lightly treated areas, where it appears softly diffused, and in areas of heavier
application, where rich, velvety patches are apparent. Piazzetta further enliv-
ened the drawing by highlighting certain areas, such as the pear and the tip
of the boy's nose, with touches of white chalk to indicate reflections of light.

GIOVANNI BATTISTA PIRANESI
Italian, 1720–1778
Study for Part of a Large Magnificent Port, ca. 1749–1750
Red and black chalk and brown and reddish wash
38.5 x 52.8 cm (15⅛ x 20¹³/₁₆ in.)
88.GB.18

Piranesi was one of the greatest draughtsmen and etchers of eighteenth-
century Italy. He is best known for his etched views of contemporary Rome and
of that city's antiquities as well as for imaginary amalgams based on these subjects. This fanciful drawing of an antique port served as a preparatory study for
an etching. The extremely complex composition is made up of massive inter-
locking architectural forms, including a portal and bridge in the foreground, a
triumphal arch in the center, and a colosseumlike structure, truncated pyramid,
and obelisk in the distance. Each section of architecture is decorated with an
elaborate sculptural program. In the foreground are boats reminiscent of those
found in Piranesi's native Venice. The entire composition is further dramatized
by the smoke billowing among the structures. The drawing's technique is as
creative as the composition. Piranesi first drew the basic structures in chalk,
then liberally added brown and red wash for shading and for many of the
details. The wash gives a vibrancy to the drawing that is not discernible in the
etching. At least two known drawings preceded this one in the process of creating the etching, which was first published in 1750 as part of the series entitled
Opere varie. The black chalk grid lines throughout were made so that the
design could be transferred to the etching plate.

PETER PAUL RUBENS
Flemish, 1577–1640
Korean Man, ca. 1617–1618
Black chalk and touches of red chalk in the face
38.4 x 23.5 cm (15 ⅛ x 9 ¼ in.)
83.GB.384

This imposing portrait of a man in formal costume, used by Rubens as the basis
for one of the central figures in a painting, is among the earliest representations
of a Korean on European soil. One of the most meticulous of Rubens' portraits,
the drawing is enriched by the use of color in the face. The fine execution of the
face contrasts with the broad rendering of the voluminous robe.

REMBRANDT VAN RIJN
Dutch, 1606–1669
Nude Woman with a Snake (as Cleopatra), ca. 1637
Red chalk and white chalk heightening
24.7 x 13.7 cm (9¹¹/₁₆ x 5⁷/₁₆ in.)
81.GB.27

This is one of the best preserved and most fluent of all of Rembrandt's red chalk drawings. He interpreted his famous subject as a living image, not at all idealized, and with great incisiveness in the characterization of her expression as she is about to kill herself by allowing the asp to bite her breast. The range of draughtsmanship in the form is quite remarkable, from the fine modeling of the right side of the figure to the virtuoso passages of freely sketched drapery at the left.

REMBRANDT VAN RIJN
Dutch, 1606–1669
Landscape with the House with the Little Tower, ca. 1651–1652
Pen and brown ink and brown wash
9.7 x 21.5 cm (3¹³/₁₆ x 8⁷/₁₆ in.)
83.GA.363

This is among the most sensitive of Rembrandt's landscape drawings and one of his most experimental. He created the suggestion of space and atmosphere in the foreground through the use of a very few lines. Then, with a complex mixture of thin lines, dotting, and a varied wash, he elaborated a rich and diverse background. The quality of air and light in this sheet is remarkable. It is a prime example of a great artist evoking a vast and varied ambience with economy of means.

NICOLAS POUSSIN
French, 1594–1665
Apollo and the Muses on Parnassus, ca. 1626–1632
Pen and brown ink and brown wash
17.6 x 24.5 cm (6^{15}/$_{16}$ x 9^{11}/$_{16}$ in.)
83.GG.345

Poussin was an admirer of Raphael and of the entire classical tradition. This drawing is based on the famous fresco by Raphael in the Stanza della Segnatura in the Vatican. The drawing is unusually animated for Poussin but shows his characteristic tendency to abstract forms and to employ wash quite broadly. He used this study in formulating a painting now in the Museo del Prado, Madrid.

ANTOINE WATTEAU
French, 1684–1721
Two Studies of a Flutist and One of the Head of a Boy, ca. 1716–1717
Red, black, and white chalk
21.4 x 33.6 cm (8⁷⁄₁₆ x 13³⁄₁₆ in.)
88.GB.3

Watteau was one of the most influential artists of the Rococo period in France.
He most often focused on lyrical subjects such as the *fête galante*, in which
themes of love, beauty, and music predominated. Watteau, who drew unceas-
ingly, probably made this drawing while attending a concert. He depicted
a flute player from two angles and, in view of the differing expressions on the
musician's face, at two distinct musical moments. The sketch at the right shows
the flutist with wrinkled brow concentrating on a seemingly difficult passage,
while the raised head and brow of the central figure suggest lighter strains. The
young boy at the left appears to be listening. Watteau became a master at the
technique of combining red, black, and white chalks, known in France as *trois
crayons*. His spontaneity in working the three chalks enhanced the liveliness
of this drawing.

JACQUES-LOUIS DAVID
French, 1748–1825
The Lictors Carrying the Bodies of the Sons of Brutus, 1787
Pen and black ink and gray wash
32.7 x 42.1 cm (12⁷⁄₈ x 16⁹⁄₁₆ in.)
84.GA.8

In this drawing David has depicted a scene from ancient Roman history that
is emblematic of great moral and patriotic virtue. Brutus, who has condemned
his sons to death for treason, sits stoically in the foreground contemplating the
tragedy while their bodies are carried away at the left and his wife and daughters
mourn at the right. The drawing is highly finished, and its composition was
employed, with changes, in David's painting of the same theme now in the
Musée du Louvre, Paris. This is among the most monumental and complete
of David's drawings.

WILLIAM BLAKE
English, 1757–1827
Satan Exulting over Eve, 1795
Graphite, pen and black ink, and watercolor over color print
42.6 x 53.5 cm (16¾ x 21¹⁄₁₆ in.)
84.GC.49

This is a powerful example of the art of Blake, inspired by both the Bible and his own miltonian approach to it. Although a fine technician, Blake preferred broadly executed compositions of this kind, intentionally avoiding the naturalism of form and space refined by Western artists before him.

FRANCISCO JOSE DE GOYA Y LUCIENTES
Spanish, 1746–1828
Contemptuous of the Insults,
ca. 1808–1812
Brush and india ink
29.5 x 18.2 cm (10¼ x 7³⁄₁₆ in.)
82.GG.96

This sheet was once part of an album of drawings that Goya seems to have made for his private use. Drawn with great tonal subtlety and fluidity, it shows a gentleman expressing disdain for the insults of gnomes, possibly a symbol of Goya's contempt for the soldiers who caused great devastation in Spain during the Napoleonic wars.

JEAN-AUGUST-DOMINIQUE INGRES
French, 1780–1867
Portrait of Lord Grantham, 1816
Graphite
40.5 x 28.2 cm (15¹⁵/₁₆ x 11⅛ in.)
82.GD.106

Lord Grantham was an Englishman who visited Rome and commissioned this drawn portrait from Ingres. It shows the subject standing in front of a distant view of Saint Peter's and exemplifies Ingres' refined manner and purity of line. At the same time, it shows great linear variety, ranging from thin, very fine strokes to bold and heavy areas. The general restraint in graphic manner is fully in accord with the portrayal of the subject.

EUGENE DELACROIX
French, 1798–1863
The Education of Achilles, ca. 1860–1862
Pastel on paper
30.6 x 41.9 cm (12¹/₁₆ x 16½ in.)
86.GG.728

Delacroix was the most important painter of the Romantic period in France. His style reflects a typically Romantic interest in themes of exoticism and heroism coupled with the colorism and free handling of paint he admired in the work of Peter Paul Rubens. The subject of this pastel, drawn from Greek mythology, is the homeric hero Achilles, shown learning the art of hunting from the centaur Chiron. The vital energy Delacroix gave Achilles and Chiron as they gallop through the landscape is matched by the vigorous style in which he made the drawing. He used pastels in a painterly fashion, applying them with a sweeping gesture for the sky and landscape and more minutely in the details of the figures and vegetation. The palette is especially rich, consisting primarily of dark green, blue, and brown punctuated by vibrant flesh tones and touches of red. Delacroix completed two paintings with the same composition. He made this pastel as a gift for his friend the writer George Sand, who had expressed her admiration for those paintings.

HONORE DAUMIER
French, 1808–1879
A Criminal Case, ca. 1860
Pen, watercolor, and gouache
38.5 x 32.8 cm (15⅛ x 12¹³/₁₆ in.)
89.GA.33

Best known for his satirical lithographs, Daumier also made drawings, water-colors, paintings, and sculpture. This watercolor comes from his later career, when his style became broad and vigorous and his works more monumental in feeling despite their relatively small scale. Here he has depicted a lawyer conferring with a worried client in a court of law before seemingly disinterested spectators; a guard stands at attention in the rear. Daumier was fascinated by the French legal system, having worked as a youth in the Parisian courts and in later years as a court illustrator. He experienced the intricacies and ironies of the judicial world firsthand when he stood trial and served a prison term for the publication of a slanderous print. This image is one of the most powerful of Daumier's many lithographs and watercolors of courtroom scenes.

VINCENT VAN GOGH
Dutch, 1853–1890
Portrait of Joseph Roulin, 1888
Reed and quill pens and brown ink and black chalk
85.GA.299

Joseph Roulin was a postal worker with whom Van Gogh became friends
during his stay in Arles in 1888–1889. Roulin and his wife and children sat for
Van Gogh many times. The artist wrote that the postman's physical appearance
and philosophy reminded him of Socrates. Van Gogh's strong impression of
Roulin is evident in the sitter's absolute frontality, close-up placement, and
intense gaze. The drawing is animated by the patterns of his clothing and beard,
which are heightened by the web of lines in the background. Van Gogh used
two different pens and various techniques of hatching, shading, and stippling
to achieve the remarkably varied textures.

PAUL CEZANNE
French, 1839–1906
Still Life, ca. 1900
Watercolor and pencil
48 x 63.1 cm (18^{15}/$_{16}$ x 24^{7}/$_{8}$ in.)
83.GC.221

This is one of the largest and most ambitious of all of Cézanne's watercolors. It is very highly worked up, and in this respect, as in its scale, it shares the character of his paintings. At the same time, the richness of technique, with levels of translucent watercolor creating both carefully structured forms and veils of colored light, brings out the possibilities of this medium to the fullest. It is noteworthy that the almost kaleidoscopic effects of color enhance the monumental purity of the blank white areas of the paper. In every respect this is one of the artist's most accomplished works in drawing.

DECORATIVE ARTS

The Museum's collection of decorative arts—of which less than a quarter is illustrated here—consists mainly of objects made in Paris from the mid-seventeenth to the end of the eighteenth century. It contains furniture, silver, ceramics, tapestries, carpets, and objects made of gilt bronze. The latter include wall lights, chandeliers, firedogs, and inkstands. All of these objects are displayed in chronological order, together with suitable paintings and sculpture, in twelve galleries, three of which are fitted with eighteenth-century paneling. Some pieces of furniture are also to be found in the paintings galleries.

J. Paul Getty began to acquire furniture in the late 1930's and continued to acquire slowly until the 1960's. Then, having opened his original small museum in Malibu, he stopped collecting for a time. At this point the entire collection of decorative arts, which consisted of about thirty objects, was housed in two galleries.

With the prospect of filling the much larger museum that was to open in 1974, Mr. Getty began to acquire again in the early 1970's and continued to do so until his death in 1976. In this period strong roots were put down, and some new curatorial policies were established. Gradually, a conception began to be realized, that of forming a representative collection of French decorative arts from the early years of the reign of Louis XIV (1643–1715) through the early decades of the nineteenth century, known as the Empire period.

With the formation of the J. Paul Getty Trust, the confines of the collection were expanded to include objects made in Germany as well as those from Italy and, more recently, Northern Europe. It is to be expected that this expansion of the decorative arts collection will take place slowly. In keeping with the standards already set, only the finest pieces will be acquired, to be gleaned from an international market that is often restricted by laws pertaining to legal exportation. The expansion of the collection will obviously outstrip the space presently provided to house it. As new objects are acquired, therefore, other objects will be removed into storage. Thus the galleries will be ever-changing, and visitors returning to the Museum will be rewarded by the sight of newly acquired objects.

READING AND WRITING TABLE
Attributed to Pierre Golle (d. 1683)
French, Paris, ca. 1670–1675
Oak veneered with ivory, horn, and ebony; gilt bronze moldings
63.5 x 48.5 x 35.5 cm (2 ft. 1 in. x 1 ft. 7⅛ in. x 1 ft. 2 in.)
83.DA.21

This small table, described in the posthumous inventory of Louis XIV, is one
of the few surviving pieces of furniture to have belonged to that great monarch.
Among the rarest and earliest pieces in the collection, it is decorated with
horn painted blue underneath, perhaps to resemble lapis lazuli. The table is
fitted with an adjustable writing surface and a drawer constructed to contain
writing equipment.

ONE OF TWO COFFERS ON STANDS

Attributed to André-Charles Boulle (1642–1732)
French, Paris, ca. 1680–1685
Oak veneered with brass, pewter, tortoiseshell, and ebony; gilt bronze mounts
156.6 x 89.9 x 55.9 cm (5 ft. 1⅛ in. x 2 ft. 11⅜ in. x 1 ft. 10 in.)
82.DA.109.1–.2

This coffer and its pendant were almost certainly made by André-Charles
Boulle at a time when he was making numerous pieces of furniture for the
Grand Dauphin, son of Louis XIV. A coffer of similar form delivered by Boulle
is listed in the Grand Dauphin's inventory of 1689. Such coffers were intended
to hold jewelry and other precious objects. The large gilt bronze straps at the
front can be let down to reveal small drawers.

TABLE

French, Paris, ca. 1680
Oak veneered with brass, pewter, tortoiseshell, ebony, and olive wood; gilded wood and gilt bronze mounts
76.7 x 42 x 36.1 cm (2 ft. 6½ in. x 1 ft. 4½ in. x 1 ft. 2¼ in.)
82.DA.34

The top of this small table folds open to reveal a scene of three women taking tea beneath a canopy. It is therefore likely that the table was intended to support a tea tray. The table is decorated with four large dolphins and four fleurs-de-lis, all in tortoiseshell. These emblems were used by the Grand Dauphin, son of Louis XIV, for whom this table and its pair, now in the British Royal Collection, probably were made. The scene on the top was copied after an engraving by the Huguenot ornamentalist and engraver Daniel Marot.

CABINET ON STAND

Attributed to André-Charles Boulle (1642–1732)
French, Paris, ca. 1675–1680
Oak veneered with pewter, brass, tortoiseshell, horn, ebony, ivory, and wood marquetry; bronze mounts; painted, gilded figures
230 x 151.2 x 66.7 cm (7 ft. 6½ in. x 4 ft. 11½ in. x 2 ft. 2¼ in.)
77.DA.1

This cabinet was probably made by André-Charles Boulle for Louis XIV, whose likeness appears in a bronze medallion above the central door, or as a royal gift. The medallion is flanked by military trophies, while the marquetry on the door below shows the cockerel of France triumphant over the eagle of the Holy Roman Empire and the lion of Spain and the Spanish Netherlands. The cabinet is supported by two large figures, the one on the right being Hercules. Clearly made to glorify Louis XIV's victories, this piece is one of a pair; the other is privately owned.

TABLE

Attributed to André-Charles Boulle (1642–1732)

French, Paris, ca. 1680

Oak veneered with brass, pewter, tortoiseshell, ebony, horn, ivory, and marquetry of stained and natural woods

72 x 110.5 x 73.6 cm (2 ft. 4⅜ in. x 3 ft. 7½ in. x 2 ft. 5 in.)

71.DA.100

The intricate marquetry of different materials is typical of the work of André-Charles Boulle. The wood marquetry flowers on the top of this table are very naturalistic and can readily be identified.

LONG CASE CLOCK

Attributed to André-Charles Boulle (1642–1732); movement made by Antoine Gaudron (d. ca. 1707)

French, Paris, ca. 1690

Oak veneered with tortoiseshell, pewter, brass, and ebony; enameled metal; gilt bronze mounts

195 x 48 x 19 cm

(6 ft. 4½ in. x 1 ft. 6½ in. x 7½ in.)

88.DA.16

This is an early example of a long case clock. The type of movement it houses was introduced by the Dutchman Huygens in 1673. It was a far more accurate timekeeper than earlier clocks.

MODEL FOR A MANTEL CLOCK

French, Paris, ca. 1700
Terracotta; enameled metal plaques
78.7 x 52.1 x 24.2 cm (2 ft. 7 in. x 1 ft. 8½ in. x 9½ in.)
72.DB.52

It is remarkable that this full-sized terracotta model for a clock has survived in such a good state of preservation. It was often the custom in the eighteenth century for a model to be ordered from a cabinetmaker for approval before the finished object was made. All of the very few such models that survive today date from the late eighteenth century. No clock of this model, which features the rape of Persephone by Pluto below the dial, seems to exist. It is possible that the model was made by one of the royal craftsmen such as the great André-Charles Boulle for the approval of Louis XIV.

MEDAL CABINET
Attributed to André-Charles Boulle (1642–1732)
French, Paris, ca. 1710–1720
Oak veneered with ebony, brass, and tortoiseshell; gilt bronze mounts;
marble top
82.5 x 140 x 72.5 cm (2 ft. 8½ in. x 4 ft. 7¼ in. x 2 ft. 4¼ in.)
84.DA.58

The pair to this cabinet, in the State Hermitage, Leningrad, retains its twenty-
four shallow drawers for medals. These cabinets are of unique form and
obviously were commissioned by a wealthy collector of ancient coins or
medals. A pair of cabinets of similar form are listed in an inventory taken at the
death of Jules-Robert de Cotte in 1767. It is possible that he had inherited the
cabinets from his illustrious father, the royal architect Robert de Cotte. The latter was closely associated with André-Charles Boulle, whose style the cabinets
represent well.

TAPESTRY
Made by Jean de la Croix (active 1662–1712)
French, Gobelins, before 1712
Wool and silk
316 x 328 cm (10 ft. 4¼ in. x 10 ft. 9 in.)
85.DD.309

This tapestry, representing the month of December and the chateau of
Monceaux, is one of a series of twelve, emblematic of the months of the year
and of French royal splendor. Each weaving portrays King Louis XIV in a
daily activity, a royal residence, and a selection of the monarch's riches and
exotic animals. The series was conceived by Charles Le Brun for production
at the royal workshop at Gobelins, and seven complete sets, woven with
gold thread, were made for the Crown between 1668 and 1711. This example,
however, is one of a number made as private commissions and bears the
weaver's name.

EWER

(Ewer) Chinese, Kangxi, 1622–1722;
(mounts) French, Paris, ca. 1700–1710
Hard-paste porcelain; polychrome
enamel decoration; gilt bronze
mounts
46.1 x 35.2 x 13.8 cm
(1 ft. 6⅛ in. x 1 ft. 1⅞ in. x 5⅜ in.)
82.DI.3

This ewer is an early example of the
Parisian fashion for fitting Oriental
porcelain with gilt bronze mounts.
Throughout the late seventeenth and
eighteenth centuries the French had a
passion for objects from the East but
were not content with the original
simple lines and elegant proportions.
They preferred to add gilt bronze or
silver mounts, in keeping with the
luxurious tastes of the nobility.

LIDDED BOWL
(Porcelain) Japanese, Imari, ca. 1700; (mounts) French, Paris, ca. 1717–1722
Hard-paste porcelain; polychrome enamel decoration; gilding; silver mounts
27.9 x 34 x diam. 27.5 cm (11 in. x 1 ft. 1⅜ in. x 10⅞ in.)
79.DI.123.a–.b

The bluish white color of Japanese porcelain was better suited to mounts of silver than it was to gilt bronze, such as those on a ewer of similar date (see p. 163).
The lid of this bowl was created by joining an inverted plate and a small lid of
similar decoration. Such assemblages are not uncommon and are a mark of the
inventiveness of eighteenth-century French craftsmen, many of whom remain
anonymous. The silver mounts do not bear any date marks, but they can be
dated stylistically to the early decades of the eighteenth century. It was during
these years that nearly all Imari porcelain mounted with silver was produced.

ONE OF A PAIR OF SCREENS
French, Savonnerie, 1714–1740
273.6 x 194.9 cm (8 ft. 11¾ in. x 6 ft. 4¾ in.)
83.DD.260.1–.2

This three-paneled screen of knotted wool pile was made in the Savonnerie
workshops, the royal manufactory that produced carpets, screens, and upholstery covers exclusively for the French Crown. Jean-Baptiste Belin de Fontenay
and Alexandre-François Desportes provided the cartoons for this, the largest
model. This example retains its original brilliant coloring.

COMMODE
Made by Etienne Doirat (ca. 1670–1732)
French, Paris, ca. 1725–1730
Oak and pine veneered with kingwood; gilt bronze mounts; marble top
86.4 x 168.9 x 71.7 cm (2 ft. 10 in. x 5 ft. 6½ in. x 2 ft. 4¼ in.)
72.DA.66

Doirat, whose name is stamped on this commode, often worked for the German market. Since the Parisian guild of *menuisiers-ébénistes* only instigated a rule in 1751 that all works be stamped with the maker's name, it is unusual to find a piece of furniture stamped at this early date.

This commode may well have been intended for a German patron who would have liked its large scale, its exaggerated *bombé* shape, and the profusion of mounts. The lower drawer is provided with a small wheel, located at the front in the center, to support its weight and to facilitate its opening.

TERRESTRIAL GLOBE

Painted decoration attributed to the
Martin brothers (active ca. 1725–1780)
French, Paris, ca. 1728–1730
Printed paper; papier-mâché;
gilt bronze; wood painted with
vernis Martin
110 x (diam.) 45 cm
(3 ft. 7 in. x 1 ft. 5½ in.)
86.DH.705.1–.2

This globe and its accompanying
celestial globe were designed and
assembled by Abbé Jean-Antoine
Nollet (1700–1770), a fashionable
scientist who taught physics to the
royal children. The terrestrial globe
bears a dedication to the duchesse
du Maine, the wife of Louis XIV's
first illegitimate child, and is dated
1728. The celestial globe is dedicated to
her nephew, Louis de Bourbon, comte
de Clermont, and is dated 1730. The
stands are painted with a yellow *vernis*
ground, polychrome flowers, and red
cartouches bearing chinoiserie scenes.
This work probably was carried out
by the Martin brothers, who obtained
a monopoly for the technique in the
mid-eighteenth century.

Many contemporary portraits show
the sitter with a globe close by, and
such objects came to be considered
essential adornments for the *biblio-
thèques* and *cabinets* of the aristocracy.

COMMODE

Made by Charles Cressent (1685–1768)
French, Paris, ca. 1735
Pine veneered with *bois satiné* and amaranth; gilt bronze mounts; marble top
90.2 x 136.5 x 64.8 cm (2 ft. 11½ in. x 4 ft. 5¾ in. x 2 ft. 1½ in.)
70.DA.82

The commode was made by Charles Cressent, a cabinetmaker who also cast
and gilded his own mounts. This practice of casting bronze in his workshop
transgressed the strict rules of the Parisian guild of *fondeurs-doreurs*, and
Cressent was often fined for his infringement. In order to pay these fines, he
held sales of his stock and wrote the descriptive catalogues himself. The catalogue to the sale of 1756 survives, and this commode is entered as number 132.

SIDE TABLE
French, Paris, ca. 1730
Carved and gilded oak; marble top
89.3 x 170.2 x 81.3 cm (2 ft. 11 in. x 5 ft. 7 in. x 2 ft. 8 in.)
79.DA.68

This table is intricately carved with lions' heads, dragons, serpents, and Chimerae, or composite mythological beasts. While the carving is deep and pierced in many areas, the table has remarkable strength and is well able to support its heavy marble top. It was originally part of a set including two smaller side tables; they would have stood in a large *salon* fitted with paneled walls carved with similar elements. It is possible that chairs, a settee, and a fire screen, all similarly carved, once completed the furnishings of the room.

WALL CLOCK
Movement made by Charles Voisin (1686–1760)
French, Paris and Chantilly, ca. 1740
Soft-paste porcelain; polychrome enamel decoration; gilt bronze mounts; enameled metal dial
74.9 x 35.6 x 11.1 cm (2 ft. 5½ in. x 1 ft. 2 in. x 4⅜ in.)
81.DB.81

The case of this clock was made of soft-paste porcelain at the Chantilly manufactory, which was established in 1725 by the prince de Condé. The prince was a great collector of Japanese porcelain, and initially the Chantilly manufactory produced wares in the Japanese style. This clock may have been intended to hang over a bed and is fitted with a repeating mechanism; the pull of a string makes it strike the time to the nearest hour and a quarter.

TAPESTRY

French, Gobelins, ca. 1730–1743
Wool and silk
355 x 262.5 cm (11 ft. 10 in. x 8 ft. 9 in.)
85.DD.100

Tapestries of this form, called *portières*, were used extensively during the seventeenth and first half of the eighteenth centuries in French royal residences, where they were hung over every door in the grand *appartements*. They traditionally were decorated with symbols of royal power. This example, though designed by Pierre-Josse Perrot for Louis XV (r. 1715–1774), incorporates symbols—such as the head of the god Apollo—that were associated with the king's great grandfather, Louis XIV, the Sun King. Twenty-eight *portières* of this model were woven, this one under the direction of Etienne-Claude Le Blond, whose name is woven in the outer border.

ONE OF A PAIR OF COMMODES

Attributed to Joachim Dietrich (d. 1753)
German, Munich, ca. 1745
Painted and gilded pine; marble top
83.2 x 126.4 x 61.9 cm (2 ft. 8¾ in. x 4 ft. 1¾ in. x 2 ft. ⅜ in.)
72.DA.63.1–.2

The design of this ornately carved German commode and its pair was influenced by the engravings of François de Cuvilliés, architect for the Elector of Bavaria and one of the leading interpreters of the Rococo style. The carved side panels specifically follow an engraving by Cuvilliés in one of almost a hundred sets of ornamental designs which he published between 1738 and 1756.

ONE OF A SET OF TWO ARMCHAIRS AND TWO SIDE CHAIRS

French, Paris, ca. 1735–1740
Carved, gessoed, and gilded
beechwood; modern silk upholstery
110.5 x 76.6 x 83.7 cm (3 ft. 7½ in. x
2 ft. 6⅛ in. x 2 ft. 8⅞ in.)
82.DA.95.1–.4

This early Rococo chair belonged to a set; two additional armchairs and two side chairs exist. The chairs are not stamped with their maker's name since they predate the guild regulations requiring this. The upholstered seat, back, and arm cushions were designed so that they could be removed and recovered according to the season.

ONE OF A PAIR OF MOUNTED CERAMIC GROUPS

(Figure, rockwork, and dog) Chinese, Kangxi, 1662–1722; (spheres) Chinese, Qianlong, 1736–1795; (flowers) French, Chantilly, ca. 1740; (mounts) French, Paris, ca. 1740–1745
Hard-paste porcelain; soft-paste porcelain; polychrome enamel decoration; gilt bronze mounts
30.4 x 22.8 x 12.7 cm
(1 ft. x 9 in. x 5 in.)
78.DI.4

As is often the case with eighteenth-century groups of mounted Oriental porcelain, this example is composed of disparate elements never intended to be used together. The boy was made to be freestanding. The rockery originally formed a base for the Fo dog, but these elements were separated and the perfume ball placed in between. The group was further embellished with French porcelain flowers and a French gilt bronze base. The picturesque ensemble would have appealed to the eighteenth-century taste for chinoiserie.

ONE OF A PAIR OF DECORATIVE BRONZES

Painted decoration attributed to the Martin brothers (active ca. 1725–1780) (Bronzes) French, Paris, 1745–1749; (mounts) French, Paris, 1738–1750
Painted bronze; silver
22.8 x 11.5 x 15.2 cm
(9 in. x 4½ in. x 6 in.)
88.DH.127.1–.2

This bronze of a boy carrying a silver bundle of sugar cane and its pair are ornamental figures, masquerading as sugar casters, that probably were not intended to be functional, given their size and unwieldiness. The painted surfaces are attributed to the Martin brothers, who specialized in this type of decoration. These unusual, inventive pieces demonstrate the French passion for chinoiserie. They originally belonged to Madame de Pompadour, mistress to Louis XV, and are described in the daily record book of the dealer Lazare Duvaux under the month of September 1752, when they were cleaned for her.

ONE OF A PAIR OF TUREENS AND STANDS

Made by Thomas Germain (1673–1748)
French, Paris, 1744–1750
Silver
(Tureen) 30 x 34.9 x 28.2 cm (11³/₁₆ in. x 1 ft. 1³/₄ in. x 11⅛ in.); (stand) 4.2 x
46.2 x 47.2 cm (1⅝ in. x 1 ft. 6³/₁₆ in. x 1 ft. 6⁹/₁₆ in.)
82.DG.13.1–.2

This large tureen and stand, along with its pair, are marked for the years 1744 to
1750. They are attributed to the royal silversmith Thomas Germain, who died
in 1748, but may have been completed by his son, François-Thomas Germain.
Both father and son were exceptionally skilled craftsmen who served many
European royal courts, especially those of France and Portugal. It is not known
who commissioned the tureens, though the tassels engraved around the later
coat of arms suggest that the original patron was a Portuguese archbishop. At
the end of the eighteenth century, the tureens passed into the possession of the
British peer Robert, first Lord Carrington, who had the coat of arms removed
and replaced with his own.

The finely worked vegetables and crustaceans on the lid, probably cast from
natural models, demonstrate the virtuoso talent of the silversmith.

COMPOUND MICROSCOPE WITH CASE

French, Paris, ca. 1750

(Microscope) gilt bronze; mirror glass; enamel; shagreen; (case) wood; tooled and gilded leather; brass closing fixtures; original silk velvet and silver braid lining

(Microscope) 48 x 28 x 20.5 cm (1 ft. 6⅞ in. x 11 in. x 8¹/₁₆ in.); (case) 66 x 34.9 x 27 cm (2 ft. 2 in. x 1 ft. 1¾ in. x 10⅝ in.)

86.DH.694

This microscope was made for a noble dilettante scientist who would have used it in his *laboratoire* or *cabinet de curiosités* to explore the mysteries of the natural world being revealed during the Age of Enlightenment.

A microscope of the same form, known to have belonged to Louis XV, was kept in his *observatoire* at the chateau of La Muette. The Museum's example is still in working condition. In a drawer in the base of the case are extra lenses and eighteenth- and nineteenth-century specimen slides of such items as geranium petals, hair, and fleas.

CARTONNIER WITH *SERRE-PAPIER* AND CLOCK

Made by Bernard van Risenburgh (d. 1765/66); clock movement made by Etienne Le Noir (1699–after 1778)

French, Paris, ca. 1746–1749

Oak veneered with ebonized wood and painted with *vernis Martin*; gilt bronze mounts; enameled metal dial; painted bronze figures

192 x 103 x 41 cm (6 ft. 3⅝ in. x 3 ft. 4⁹/₁₆ in. x 1 ft. 4⅛ in.)

83.DA.280

The piece is stamped B.V.R.B. for the cabinetmaker Bernard van Risenburgh. The pigeonholes in the central section and the narrow cupboards at the sides of the base were intended to contain papers. The black and gold decoration is of *vernis Martin*, a French imitation of Oriental lacquer named after the Martin brothers who invented it. In the Rococo period the use of Oriental lacquer panels to decorate expensive furniture was popular. However, the exotic lacquer was expensive for the client or difficult to obtain, so Parisian craftsmen learned to imitate it. The clock movement is dated 1746.

BUST OF LOUIS XV OF FRANCE

French, Mennecy, ca. 1750–1755
Soft-paste porcelain
43.2 x 24.5 x 14.5 cm
(1 ft. 5 in. x 9 9/16 in. x 5 11/16 in.)
84.DE.46

In 1734 the duc de Villeroy established a soft-paste porcelain manufactory, which after 1748 was located at Mennecy. Because of the difficulties of firing large pieces of porcelain in the kilns, this bust of the king was made in two pieces, joined above the crown on the plinth. The plinth is decorated with an asymmetrical Rococo cartouche (enclosing the French royal coat of arms) and with military trophies.

DOUBLE DESK

Made by Bernard van Risenburgh (d. 1765/66)
French, Paris, ca. 1750
Oak veneered with tulipwood and kingwood; gilt bronze mounts
107.8 x 158.7 x 84.7 cm (3 ft. 6½ in. x 5 ft. 2½ in. x 2 ft. 9⅜ in.)
70.DA.87

The form of this massive desk, stamped B.V.R.B. (see p. 175), is unique. Flaps let down on both sides to form writing surfaces, revealing pigeonholes and drawers. The desk was bought in Paris by the Duchess of Hamilton in the 1760's and passed by inheritance through her second marriage to the Duke of Argyll, in whose family it remained until it was acquired by J. Paul Getty.

ONE OF A PAIR OF WALL LIGHTS

French, Paris, 1745–1749
Gilt bronze
72.4 x 47.5 x 26.7 cm
(2 ft. 4½ in. x 1 ft. 6¾ in. x 10½ in.)
89.DF.26.1–.2

Boldly sculptural, these wall lights are massive in scale and form. They undoubtedly were designed for a formal interior, probably en suite with other wall lights and a chandelier. Each light bears the tax stamp of the crowned "c," indicating that they were made between 1745 and 1749. The name of the maker is not known.

COMMODE

Made by Jean-Pierre Latz (ca. 1691–1754)
French, Paris, ca. 1745–1749
Oak veneered with *bois satiné*; gilt bronze mounts; marble top
87.7 x 151.5 x 65 cm (2 ft. 10½ in. x 4 ft. 11⅝ in. x 2 ft. 2⅝ in.)
83.DA.356

Although this commode is not stamped with a cabinetmaker's name, it can be firmly attributed to Latz since his stamp is found on a commode of precisely the same design now in the Palazzo Quirinale, Rome. That commode was taken to Italy in 1753 by Louise-Elisabeth, a daughter of Louis XV, who had married Philip, Duke of Parma, son of Philip V of Spain (r. 1700–1724, 1724–1746).

COMMODE
Attributed to Joseph Baumhauer (d. 1772)
French, Paris, ca. 1750
Oak set with Japanese lacquer and painted with *vernis Martin*; gilt bronze
mounts; marble top
88.3 x 146.1 x 62.6 cm (2 ft. 10¾ in. x 4 ft. 9½ in. x 2 ft. ⅝ in.)
55.DA.2

The front and sides of this commode are set with panels of Japanese lacquer,
the seams of which are hidden under the mounts. The remaining surfaces are
painted in imitation of *nashiji*, a clear lacquer sprinkled with gold. This commode
bears the label of the eighteenth-century furniture dealer Charles Darnault.

CORNER CUPBOARD
Made by Jacques Dubois (ca. 1693–1763); clock movement made by Etienne
Le Noir (1699–after 1778)
French, Paris, ca. 1744
Oak veneered with *bois satiné*, tulipwood, and rosewood; enameled metal;
gilt bronze mounts
289.5 x 129.5 x 72 cm (9 ft. 6 in. x 4 ft. 3 in. x 2 ft. 4½ in.)
79.DA.66

This cupboard stamped I. DUBOIS was made for Count Jan Clemens Branicki,
the head of the Polish army. It was delivered to him in Warsaw ca. 1752, and
an inventory shows that it stood in a grand *salon* as a pendant to a large stove. Of
unique form, the cupboard is based on a print of a drawing by the great Rococo
ornamentalist Nicolas Pineau. The clock movement is dated 1744.

ONE OF A PAIR OF CABINETS

Made by Bernard van Risenburgh
(d. 1765/66)
French, Paris, ca. 1750–1755
Oak veneered with *bois satiné*,
kingwood, and cherry; gilt bronze
mounts
149 x 101 x 48.3 cm (4 ft. 10⅝ in. x
3 ft. 3¼ in. x 1 ft. 7 in.)
84.DA.24.1-.2

Stamped B.V.R.B. (see p. 175), this
low cabinet and its pair are of unique
form and were probably used to hold
small objets d'art such as porcelains
and bronzes. The cabinets are too
deep to use for books. A slide, situated
beneath the upper doors, can be pulled
out and may have been used by the
owner while rearranging and studying
his or her collection.

ONE OF A SET OF FOUR WALL LIGHTS

Made by François-Thomas Germain
(1726–1791)
French, Paris, 1756
Gilt bronze
99.6 x 63.2 x 41 cm (3 ft. 3¼ in. x
2 ft. ⅞ in. x 1 ft. 4⅛ in.)
81.DF.96.1-.4

Signed by François-Thomas Germain,
silversmith to the king (see p. 173),
these large, late Rococo wall lights
were finished with an unusually high
degree of attention to their casting
and gilding. The lights were commissioned by the duc d'Orléans for his
residence in Paris, the Palais Royal.
Later in the eighteenth century they
were bought by Louis XVI (r. 1774–
1792) and hung in rooms used by
Marie-Antoinette in the chateau of
Compiègne.

BASKET
French, Sèvres, 1756
Soft-paste porcelain; green ground color; gilding
22 x 20.1 x 18 cm (8⅝ x 7⅞ x 7⅛ in.)
82.DE.92

The "D" painted on the bottom of this basket indicates that it was made in 1756, the year the French royal porcelain manufactory moved from Vincennes to a newly built factory at Sèvres, a village on the Seine near Paris. The piece also bears the incised mark *PZ* for the modeler who cut the intricately pierced walls and made the Rococo ribbons entwining the handle. The green ground color with which this basket is painted was first developed in 1756, so this is an early example of its use.

The piece has an unusual shape. It, or one very similar to it, is described in the 1757 records of the manufactory as having been presented as a gift to François Boucher, the court painter who supplied many designs to the Sèvres manufactory.

WRITING AND TOILET TABLE
Made by Jean-François Oeben (ca. 1720–1763)
French, Paris, ca. 1750–1755
Oak veneered with burl ash, holly, tulipwood, and other stained and natural
exotic woods; leather; silk; gilt bronze mounts
71.1 x 80 x 42.8 cm (2 ft. 4 in. x 2 ft. 7 ½ in. x 1 ft. 4 ⅞ in.)
71.DA.103

The top of this table of multiple uses slides back, and a drawer unit can be
pulled forward, revealing another sliding top concealing compartments lined
with blue watered silk. These elaborate mechanisms and the fine floral and
trellised marquetry decorating it are trademarks of the table's maker, royal
ébéniste Jean-François Oeben. Toilet pots and writing materials would have
been kept in this table and its surfaces used for writing.

LIDDED POTPOURRI VASE

French, Sèvres, ca. 1760
Soft-paste porcelain; polychrome enamel decoration; gilding
37.5 x 34.8 x 17.4 cm (1 ft. 2 $^{3}/_{4}$ in. x 1 ft. 1 $^{11}/_{16}$ in. x 6 $^{13}/_{16}$ in.)
75.DE.11

This *vase vaisseau à mât*, or boat-shaped vase, was designed to contain scented potpourri—dried flower petals and herbs—and its lid is pierced to allow the scent to permeate the air. The painted genre scene on the front is surrounded by a ground of pink and green. The gold fleurs-de-lis on the pennant draped around the "mast" suggest royal ownership. This soft-paste porcelain vase is one of the most rare and elaborate of Rococo models produced in the French royal porcelain manufactory at Sèvres. Such vases were intended to be sold with others of various shapes to form a matching set, or *garniture*.

ONE OF A PAIR OF VASES
French, Sèvres, ca. 1760
Soft-paste porcelain; polychrome enamel decoration; gilding
29.8 x 16.5 x 14.6 cm (11¾ x 6½ x 5¾ in.)
78.DE.358.1–.2

The lower sections of this vase and its pair were designed to hold small flower-ing bulbs, while the tall central section was intended to hold potpourri, the scent of which would emanate through the pierced trellis at the top. Such vases would have formed a *garniture* with others of various shapes (see p. 193). The chinoiserie scene is rare on Sèvres porcelain, and it is even more rare to find three ground colors (pink, green, and dark blue) on one piece. Each color required a separate firing of the fragile porcelain. From archival records it is known that these vases come from a set purchased by Madame de Pompadour, mistress of Louis XV, from the Sèvres manufactory in 1760.

READING AND WRITING STAND

Made by Abraham Roentgen
(1711–1793)
German, Neuwied-am-Rhein,
ca. 1760
Pine and walnut veneered with
rosewood, walnut, ivory, ebony, and
mother-of-pearl
76.8 x 71.7 x 48.8 cm
(2 ft. 6½ in. x 2 ft. 4½ in. x 1 ft. 7¼ in.)
85.DA.216

The top of this stand is inlaid with the
coat of arms of Johann Philipp von
Walderdorff, Prince Archbishop and
Elector of Trier. Walderdorff was
Roentgen's most important patron,
and in the 1750's and 1760's the latter delivered some twenty pieces of elaborate
furniture to Walderdorff's palace. This stand is fitted with numerous concealed
drawers that open at the press of a button or turn of a key.

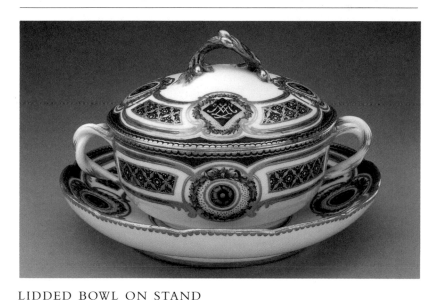

LIDDED BOWL ON STAND

Painted by Pierre-Antoine Méreaud *père* (ca. 1735–1791)
French, Sèvres, 1764
Soft-paste porcelain; polychrome enamel decoration; gilding
12.7 x 19.7 x 21.1 cm (5 x 7¾ x 8⁵⁄₁₆ in.)
78.DE.65

Decorated in the restrained transitional style of the 1760's, when the Rococo
was giving way to the newly fashionable Neoclassical taste, this lidded bowl is
painted with the monogram and coat of arms of the eighth daughter of King
Louis XV of France, Madame Louise (see p. 189). The porcelain is painted with
the mark of the painter Méreaud *père* and the date letter "ʟ" for 1764, the year
Madame Louise is recorded as having bought it.

MANTEL CLOCK

Case attributed to Etienne
Martincourt (d. after 1789);
movement made by
Charles Le Roy (1726–1779)
French, Paris, ca. 1765
Gilt bronze; enameled metal
71.1 x 59.3 x 33.3 cm (2 ft. 4 in. x
1 ft. 11⅜ in. x 1 ft. 1⅛ in.)
73.DB.78

Known from documents to have
belonged to Louis XVI, this clock is
a remarkable work of casting, since
details like the rosettes in the trellis
were cast together with the elements
they decorate. The usual practice was
to cast all of these pieces separately.
Although the clock, which came from the Palais des Tuileries, does not bear the
name of its maker, a drawing for it by the *bronzier* Etienne Martincourt has
recently been discovered. The female figures flanking the urn represent Astron-
omy and Geography.

METAL TABLE

After designs by Victor Louis (1731–1807)
French, Paris, ca. 1765–1770
Silvered and gilt bronze; *bleu turquin* marble top
78 x 129.5 x 49 cm (2 ft. 2½ in. x 4 ft. 3 in. x 1 ft. 7¼ in.)
88.DF.118

The design of this table follows two drawings—preserved in the University
Library, Warsaw—one of which was made by the French architect Victor Louis.
Louis had been commissioned in 1765 to renovate the royal palace in Warsaw for
Stanislas II Augustus, King of Poland (r. 1764–1795). A table of this design was
delivered to the palace.

CABINET

Made by Joseph Baumhauer (d. 1772)
French, Paris, ca. 1765
Oak veneered with ebony and amaranth; gilt bronze mounts; Japanese lacquer
panels; jasper top
89.6 x 120.2 x 58.6 cm (2 ft. 11¼ in. x 3 ft. 11⅜ in. x 1 ft. 11⅛ in.)
79.DA.58

The design of this early Neoclassical cabinet is severely architectonic and
includes such classical elements as fluted, canted pilasters and Ionic capitals.
Although fairly simple in appearance, it is made from rare and expensive materials. The *kijimakie* panels, of lacquer sanded so that the grain of the wood
shows, are of very fine quality and of late seventeenth-century date. A
semiprecious hard stone, yellow jasper, was used for the top instead of less
expensive marble.

The cabinet is stamped JOSEPH, the mark of the cabinetmaker Joseph
Baumhauer.

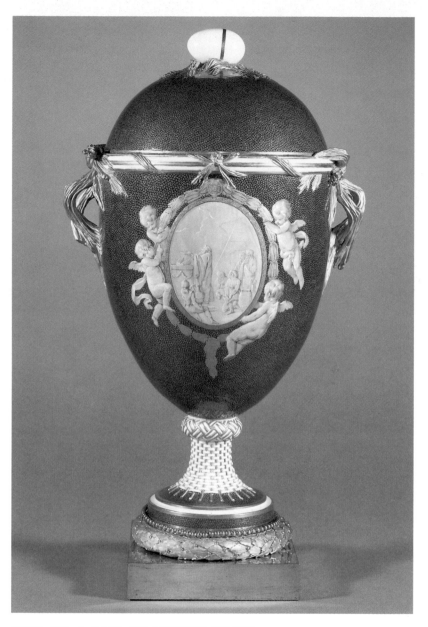

ONE OF A PAIR OF LIDDED VASES

Painted reserves attributed to Jean-Baptiste-Etienne Genest (active 1752–1789)
French, Sèvres, ca. 1769
Soft-paste porcelain; blue (*bleu Fallot*) ground color; *grisaille* enamel decoration;
gilding; gilt bronze mounts
45.1 x 24.1 x 19.1 cm (1 ft. 5 ¾ in. x 9 ½ in. x 7 ½ in.)
86.DE.520

This vase is of almost unique form. The grisaille reserves and the pale blue
ground strewn with gilded dots create a subtle finish markedly different from
the strong, brightly colored decoration frequently employed at Sèvres in the
later eighteenth century.

COMMODE

Made by Gilles Joubert (1689–1775)
French, Paris, 1769
Oak veneered with kingwood, tulipwood, holly, or boxwood, and ebony; gilt bronze mounts; marble top
93 x 181 x 67.3 cm (3 ft. ⅝ in. x 5 ft. 11¼ in. x 2 ft. 2½ in.)
55.DA.5

This commode and its pair (location unknown) were delivered in 1769 to Versailles for use in the bedchamber of Madame Louise, daughter of Louis XV (see p. 185). It is inscribed on the back with the number 2556, identifying it in the royal inventories. Gilles Joubert, best known for his Rococo work, ventured to design this commode in the newly fashionable Neoclassical style.

ONE OF A SET OF SIX WALL LIGHTS

Made by Philippe Caffiéri (1714–1774)
French, Paris, ca. 1766
Gilt bronze
64.8 x 41.9 x 31.1 cm (2 ft. 1½ in. x 1 ft. 4½ in. x 1 ft. ¼ in.)
78.DF.263.1–.4; 82.DF.35.1–.2

Signed gilt bronze objects like this one are rare since the craftsmen who made them only occasionally put their names on their work. This light is further documented by the recent discovery of a drawing signed by Philippe Caffiéri and depicting the same model. These wall lights are after a design, possibly by Caffiéri himself, made in 1768 after the redecoration of one of the residences of the king of Poland.

ONE OF A SET OF FOUR ARMCHAIRS AND ONE SETTEE

Made by Jean-Baptiste Tilliard *fils* (d. 1797)

French, Paris, ca. 1770–1775

Carved and gilded beech; modern silk velvet upholstery

101.6 x 73.6 x 74.9 cm (3 ft. 3 in. x 2 ft. 5 in. x 2 ft. 5½ in.)

78.DA.99.1–.5

This armchair originally belonged to a large suite of seating furniture (another pair of chairs from the set is in the Cleveland Museum of Art). The design of the suite suggests that it was made to furnish a *grand salon*. The carving of the Neoclassical decorative elements—columnlike legs, acanthus leaves, and egg-and-dart border—is extremely fine, although the frames of the pieces are massive.

SECRETAIRE

Made by Philippe-Claude Montigny (1734–1800)

French, Paris, ca. 1770–1775

Oak and pine veneered with tortoiseshell, brass, and pewter with ebony bandings; gilt bronze mounts

141.5 x 84.5 x 40.3 cm (4 ft. 7½ in. x 2 ft. 9 in. x 1 ft. 3¾ in.)

85.DA.378

The front and sides of this Neoclassical *secrétaire-à-abattant* are covered with large marquetry panels of tortoiseshell, brass, and pewter dating from the late seventeenth century. They originally must have been tabletops that were reused by Montigny, who was well known as a specialist in this type of work. The reuse of such old panels or making of new ones was fashionable during the late eighteenth century among the most advanced connoisseurs. This vogue is known as the "boulle revival" after the technique's greatest seventeenth-century practitioner, André-Charles Boulle.

It is unusual to find an eighteenth-century provenance for nonroyal furniture. Two previous owners—both courtiers at Versailles—have been identified for this *secrétaire*.

ONE OF A SET OF FOUR TAPESTRIES

French, Gobelins, 1772/73
Wool and silk
371 x 421 cm (12 ft. 2 in. x 13 ft. 9 ¾ in.)
82.DD.66–.69

Entitled *The Arrival of Sancho on the Island of Barataria*, this tapestry is from a set of four depicting various scenes from Cervantes' *Don Quixote* that was woven at the Gobelins manufactory. They make up one of many similar sets woven between 1763 and 1787 after cartoons by the painter Charles Coypel. The designs of the elaborate frames and surrounds were altered periodically as styles changed. Such tapestries were often presented by the French king to royalty and nobility from other countries in Europe. This set was given by Louis XVI to the Duke and Duchess of Saxe-Teschen, who were traveling in France in 1786 (the duchess was the sister of Marie-Antoinette).

A number of examples from the *Don Quixote* series exist, but the Museum's tapestries are unusual in that the colors remain relatively unfaded and retain many of the softer hues now lost on other examples.

SECRETAIRE
Made by Martin Carlin (ca. 1730–1785)
French, Paris and Sèvres, ca. 1776–1777
Oak veneered with tulipwood, amaranth, and satinwood; soft-paste porcelain
plaques with polychrome enamel decoration and gilding; enameled metal; gilt
bronze mounts; marble top
107.9 x 101 x 35.5 cm (3 ft. 6¼ in. x 3 ft. 3¾ in. x 1 ft. 2 in.)
81.DA.80

The upright *secrétaire*, a type of writing desk, came into fashion in the mid-
eighteenth century. The fall front of this example lowers to form a writing surface, revealing drawers and pigeonholes. This Neoclassical *secrétaire* is decorated
with flower-painted Sèvres porcelain plaques, an expensive fashion introduced
by the Parisian furniture dealers in the 1770's. Its unusually small size suggests
that the desk was made for a bedroom.

THREE LIDDED VASES

Models designed by Jacques-François Deparis (1746–1797); at least one vase
finished by Etienne-Henry Bono (active 1754–1781); reserves painted by
Antoine Caton (active 1749–1798); gilding by Etienne-Henry Le Guay
(1719/20–ca. 1799); jeweling by Philippe Parpette (active 1755–1806)
French, Sèvres, 1781
Soft-paste porcelain; blue (*bleu nouveau*) ground color; polychrome enamel
decoration; enamel imitating jewels; gilding; gold foil
40.8 x 24.8 x 18.4 cm (1 ft. 4 in. x 9 ¾ in. x 7 ¼ in.); 49.6 x 27.7 x 19.3 cm
(1 ft. 6 in. x 10 ⅞ in. x 7 ⅝ in.); 40.5 x 25.4 x 18 cm (1 ft. 3 ¹⁵/₁₆ in. x 10 in. x 7 ³/₁₆ in.)
84.DE.718.1–.3

This set of lavishly decorated vases (along with a similar pair now in the Walters
Art Gallery, Baltimore) formed a *garniture de cheminée* which was purchased by
Louis XVI on November 2, 1781, and placed in his library at Versailles. These
vases, prime examples of the mature Neoclassical style at Sèvres, are among the
most fully documented works of porcelain in the Museum's collection. Known
as a *vase des âges*, this shape was produced at Sèvres from 1778 on in three sizes
that differ in the rendering of the "handles": bearded male heads for the largest
size, female heads for the middle size, and boys' heads for the small size (in
Baltimore). The oval reserves on the front of each vase were painted in polychrome enamels with scenes based on engravings by Jean-Baptiste Tilliard after
designs by Charles Monnet. These engravings illustrated an edition of one of
Louis XVI's favorite books, *Aventures de Télémaque*.

   These vases are among the largest pieces of jeweled porcelain made at Sèvres.
The procedure was exceptionally expensive and demanded great skill. It is clear
that only the factory's most experienced and accomplished decorators were
involved in producing this *garniture*. No directly comparable sets of vases are
known.

ROLL-TOP DESK
Made by David Roentgen (1743–1807)
German, Neuwied-am-Rhein, ca. 1785
Veneered with mahogany; gilt bronze mounts
168.3 x 155.9 x 89.3 cm (5 ft. 6¼ in. x 5 ft. 1⅜ in. x 2 ft. 11½ in.)
72.DA.47

Made in the Neoclassical style, this desk is one of several made by Roentgen, many of which were purchased by Catherine the Great. When the writing surface is pulled out, a mechanism automatically withdraws the solid roll top back into the carcass, displaying the drawers and pigeonholes inside the desk. It is also fitted with both a reading stand and a writing-surface unit, which could be used by a person standing at the closed desk. This unit is concealed, folded, behind the gilt bronze plaque; it projects forward when the weight-driven mechanism is activated by the turn of a key. Roentgen specialized in furniture with such elaborate mechanical fittings.

STANDING VASE
Attributed to Pierre-Philippe Thomire (1751–1843)
(Porcelain) Chinese, ca. 1750; (mounts) French, Paris, ca. 1785
Hard-paste porcelain; monochrome enamel decoration; gilt bronze mounts; marble
80.7 x 56.5 cm (2 ft. 7¾ in. x 1 ft. 10¼ in.)
70.DI.115

It is possible that this large standing vase served as a jardiniere that might have been placed on a table or stand. Another vase of identical form is in the British Royal Collection; it was purchased by the Prince Regent (later George IV [r. 1820–1830]) from Thomire et Cie., the company of the famous *bronzier* Pierre-Philippe Thomire. The mounts on this vase are therefore attributed to him. It was reputedly bought by Princess Isabella Lubomirska, cousin of King Stanislas of Poland (r. 1704–1709, 1733–1735), after the French Revolution.

ONE OF A PAIR OF CANDELABRA

Attributed to Pierre-Philippe Thomire (1751–1843) after designs by Louis-Simon Boizot (?) (1743–1809)
French, Paris, ca. 1785
Gilt and patinated bronze; white and griotte marbles
82.2 x diam. 29.2 cm (2 ft. 10 ¾ in. x 11½ in.)
86.DF.521

Though the name of the maker is unknown, this pair of candelabra has been attributed on stylistic grounds to the greatest *bronzier* of the late eighteenth and early nineteenth centuries, Pierre-Philippe Thomire. A drawing in the Musée des Arts Décoratifs, Paris, shows one of the candelabra displayed on a fireplace with other objects, including a clock with vestal virgins and a small bronze figure of a reading girl on a lamp base. The clock always has been attributed to Thomire as has the figure, which was designed by Louis-Simon Boizot in 1780 for the Sèvres manufactory. It is known that Thomire did not personally produce designs for his work, calling instead on the best-known designers of his period. It therefore is possible that Boizot was responsible for the design of this candelabrum.

WINE-BOTTLE COOLER
French, Sèvres, ca. 1790
Enameled and gilded soft-paste porcelain
18.9 x diam. 25.8 cm (7 7/16 in. x 10 3/16 in.)
82.DE.5

This wine-bottle cooler is painted with a dark blue ground color and two elaborate scenes of mythological subjects surrounded by carefully tooled gilding. The scene illustrated, painted by Charles-Eloi Asselin, shows Hercules performing one of his twelve labors: capturing the man-eating horses of Diomedes. The cooler comes from one of the most important dinner services produced at Sèvres in the eighteenth century, which was ordered by Louis XVI in 1783 but not completed before his execution in 1793. The king's death marked the end of production of the costly service, of which 197 pieces had been completed. The partial service was dispersed at the time of the French Revolution. A major portion was acquired by the Prince Regent (later George IV) and can be seen today at Windsor Castle.

CABINET
Made by Guillaume Benneman (d. 1811)
(Cabinet) French, Paris, 1788; (plaques) seventeenth and eighteenth centuries
Oak veneered with ebony; *pietra dura* plaques; gilt bronze mounts; *bleu turquin* marble top
91.3 x 165.4 x 64.1 cm (2 ft. 11 15/16 in. x 5 ft. 5 1/8 in. x 2 ft. 1 1/4 in.)
78.DA.361

This cabinet was created for the bedroom of Louis XVI at the chateau of Saint-Cloud. The *pietra dura* plaques replace the original lacquer panels, which were taken off sometime after the Revolution. One of a pair, the other cabinet is now in the royal palace in Madrid.

SCULPTURE & WORKS OF ART

The Department of Sculpture and Works of Art was established in 1984. Its primary goal is to build a collection of European sculpture representing the period from the Middle Ages to the end of the nineteenth century. Its other aim is to complement the Department of Decorative Arts (responsible for Northern European decorative arts from 1650 to 1900) and to build the Museum's collection of all European decorative arts from the period prior to 1650 and in Southern Europe from 1650 to 1900.

The department now has important holdings in Renaissance ceramics and glass, a small number of silver, gold, and other precious metal objects, single examples of early seventeenth-century German and Netherlandish furniture, and Italian furniture ranging from the sixteenth through the nineteenth century. Covering the same time span, the sculpture collection includes important works by Antico, Cellini, Giambologna, Bernini, Verhulst, Girardon, Houdon, Clodion, Barye, and Carpeaux.

TWO-HANDLED "OAK-LEAF" DRUG JAR
Made by the factory of Piero di Mazeo (?) (b. 1377/87)
Italian, Florence, ca. 1420–1440
Tin-glazed earthenware
31.1 x 29.8 x diam. (lip) 14.3 cm (12¼ x 11¾ x 5⅝ in.)
85.DE.56

This large drug container, known as an *orciuolo biansato*, has an exceptionally bold
and rare shape. It was produced for the pharmacy of the Sienese hospital of
Santa Maria della Scala, whose emblem—a three-runged ladder surmounted by
a cross—appears on either side. This emblem is framed in one instance by two
birds, possibly peacocks, and in the other by two human-faced birds, or harpies, interspersed with so-called oak-leaf decoration. The latter was popular
in and around Florence in the early fifteenth century. As was also common in
Tuscany, the decoration is painted in a very thick cobalt-blue impasto often
called relief blue because it was applied so thickly that it appears to stand out in
relief. The "oak leaves" are outlined in, and scattered with, touches of manga-
nese purple. Under each handle is a "P," possibly intertwined with a backward
"C," which may be the mark of the Florentine workshop of Piero di Mazeo,
known to have been active at the time the jar was made.

HISPANO-MORESQUE DEEP DISH
Spanish, Valencia, mid-fifteenth century
Tin-glazed and lustered earthenware
10.8 x diam. 49.5 cm (4½ x 19½ in.)
85.DE.441

In the first half of the fifteenth century Moorish potters in Spain produced the finest pottery in Europe. They specialized in highly prized lusterwares like this dish that were exported throughout the Mediterranean. The difficult luster technique—which produced a shimmering, iridescent effect—was accomplished by firing metal oxides onto ceramic objects in a special "reduction" kiln starved of oxygen.

The center of the dish—known as a *brasero*—is inscribed "IHS" (Jesus Hominum Salvator), the monogram Saint Bernard of Siena held up for veneration at the end of his sermons. After his canonization in 1450, the monogram began appearing on works of art, substantiating the dating of this piece to mid-century. The leaf-spray embellishment that extends around the rim and down the sides is bryony (a tendril-bearing vine), parsley, or small carnation leaves. This motif is mainly found on Hispano-Moresque wares of the second and third quarters of the fifteenth century.

Braseros often were used as serving trenchers, although its large scale, elaborate decoration, and excellent state of preservation suggest that this deep dish was intended for display, perhaps on a credenza.

CYLINDRICAL DRUG JAR
Italian, Faenza, ca. 1480
Tin-glazed earthenware
31.5 x diam. (lip) 11.1 x diam. (max.)
12.4 cm (12 3/8 x 4 3/8 x 4 7/8 in.)
84.DE.104

This pharmaceutical *albarello* is decorated with a banderole label bordered
with scrolling leaves. The interior is
covered with a lead glaze, less expensive than the brilliant tin glazes on the
exterior. The label identifies the original contents as *syrupus acetositatis
citriorum*, or syrup of lemon juice,
which was used to reduce inflamma-
tions of the viscera, calm fevers,
quench thirst, and counteract drunk-
enness and dizziness. The vessel's
waisted form is appropriate to its use,
since it was easy to remove from a
pharmacy's crowded shelf full of similar containers. The elegant shape and
masterful glaze painting make this
piece one of the finest fifteenth-
century *albarelli* known.

EWER
Italian, Venice, ca. 1500
Free-blown soda glass; enameled and
gilt decoration
H: 27.9 cm (11 in.)
84.DK.512

Originally used as liturgical vessels in
metal, by the fifteenth century ewers
were being produced in glass and used
as table pitchers to serve wine and
other liquids. This example was
formed in four separate parts: spout
and handle (originally gilt), and body
and base. By the beginning of the sixteenth century the enameled scale
pattern made of small dots had
become a popular Venetian glass
motif. The flame design below the
neck was less common. This piece is
one of only a dozen glass ewers that
have survived intact from the period.

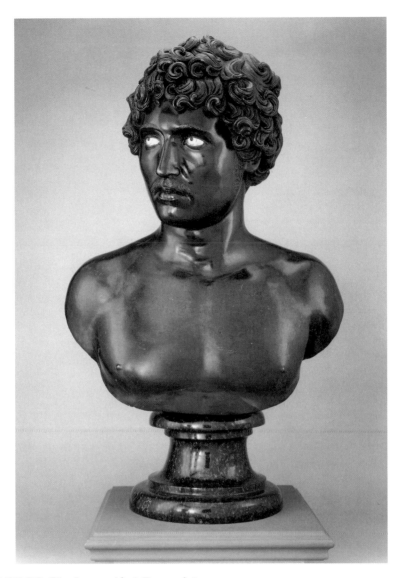

ANTICO (Pier Jacopo Alari-Bonacolsi)
Italian, ca. 1460–1528
Bust of Marcus Aurelius, ca. 1520
Bronze inlaid with silver
54.7 x 45 x 22.3 cm (21½ x 17¾ x 8¾ in.)
86.SB.688

Originally trained as a goldsmith, Pier Jacopo became the principal sculptor
at the court of Mantua in the late fifteenth and early sixteenth centuries. He
executed many bronze reductions and variants of famous antiquities, earning
himself the nickname Antico. The Museum's bronze is one of only seven known
busts generally accepted as being by him. It derives from an ancient marble
bust, now in the Hispanic Society of America, New York, which is believed to
represent the young Roman emperor Marcus Aurelius (r. A.D. 161–180). The
use of silver inlay for the eyes emulates a frequent antique practice.

BOWL

Italian, Venice, ca. 1500
Chalcedony glass
12.5 x diam. 19.6 cm (4¹⁵⁄₁₆ x 7¾ in.)
84.DK.660

This rare object is one of a small extant group of Renaissance bowls in chalcedony, or agate, glass, so called because of the opaque, marbled appearance which resembles natural hard stones. Chalcedony glass probably was developed as a result of the experiments of late fifteenth-century chemists and glass technicians such as the Venetian Barovier family. Highly prized for the beauty of its colors and as a curiosity, chalcedony glass is another example of the Renaissance revival of antiquity, since this type of glass was first made in Roman times.

HERCULES PENDANT

French, Paris, ca. 1540
Gold, enamel, and baroque pearl
6 x 5.4 cm (2⅜ x 2⅛ in.)
85.SE.237

A masterpiece of Renaissance jewelry, this pendant can be associated with the court of Francis I (r. 1515–1547) at Fontainebleau. The subject, Hercules with the columns of Cadiz, was one of many such Herculean images utilized by the king in royal commissions. The pendant's style and unusual sculptural quality recall the work of the Italian Benvenuto Cellini, who was employed at Fontainebleau from 1540 to 1545 (see p. 205).

BENVENUTO CELLINI
Italian, 1500–1571
Satyr, ca. 1542
Bronze
H: 57 cm (22 $^7/_{16}$ in.)
85.SB.69

In 1540 Cellini, a leading sixteenth-century sculptor, traveled to Fontainebleau to work for King Francis I. One of the artist's major commissions was for the Porte Dorée, the monumental palace entrance ordered by the king in 1542. The project called for a bronze lunette depicting the nymph thought to reside in the forest of Fontainebleau, to be supported by two menacing satyrs on either side of the door, with winged personifications of victory in the spandrels. Although the doorway never was completed, the Museum's bronze was cast from Cellini's wax model for one of the satyrs.

PILGRIM FLASK
Made by the Medici porcelain factory
Italian, Florence, ca. 1575–1587
Soft-paste porcelain; blue underglaze decoration
26.4 x 20 x diam. (lip) 4 cm
(10 $^1/_8$ x 7 $^7/_8$ x 1 $^9/_{16}$ in.)
86.DE.630

One of the exceedingly scarce wares produced in the Medici factory under the patronage of Francesco I, this flask is among the earliest examples of porcelain produced in Europe. Although Medici wares often display signs of their experimental nature, this work is an exceptionally success-ful, beautiful piece with unusually brilliant color.

Almost certainly a display piece, this flask reflects the influence of contem-porary maiolica and metalwork—in its shape and molded side loops—and of Chinese porcelain and Turkish Iznik ware—in its stylized floral decoration.

BASIN WITH DEUCALION AND PYRRHA
Made by Orazio Fontana (1510–1571) or his workshop
Italian, Urbino, ca. 1565–1571
Tin-glazed earthenware
6.3 x diam. 46.3 cm (2½ x 18½ in.)
86.DE.539

One of the most active, innovative ceramists of his time, Orazio helped develop a new genre of maiolica decoration with elegant grotesque motifs. The central medallion tells the story of a husband and wife who renewed the human race after a devastating flood by casting stones behind them that assumed human form (Ovid, *Metamorphoses* 1.260–415).

GIAMBOLOGNA (Jean Boulogne)
Flemish, active in Florence, 1529–1608
Bathsheba, ca. 1559
Marble
H: 115 cm (45¼ in.)
82.SA.37

Giambologna was one of the most innovative sculptors to experiment with the *figura serpentinata*, or serpentine figure, the upwardly spiraling movement that demands to be looked at from every point of view. Bathsheba's posed figure conforms to an ideal, artificial spiral achieved through the graceful but complicated twisting of limbs. For Giambologna, the naturalness of the figure's stance and her specific identity were less important than the contrast between the elegant design of her long, smooth body and the detailed drapery folds, armband, and coiffure.

COVERED *STANGENGLAS*

(Glass) Italian, Venice, ca. 1575–1600;
(mounts) made by Mathaeus Walbaum
(German, active 1582–1630/32)
Free-blown glass with embedded
canes and air bubbles; gilt silver
mounts
H: 31.3 cm (12¼ in.)
84.DK.513.1–.2

This *Stangenglas*, or tall, cylindrical,
footed drinking glass, displays the
remarkable technique of trapping
small bubbles between threadlike
white canes of glass in a net pattern.
The fine quality of this so-called
vetro a reticello is consistent with
Venetian production of the late sixteenth century. Since the mounts date
slightly later and bear the marks of
an Augsburg silversmith, the vessel
itself is likely to have been made for
export. The finial at the top represents
the Old Testament heroine Judith
holding a sword in one hand and the
head of Holofernes in the other.

EWER AND BASIN

Made by Abraham I Pfleger
(active 1558–d. 1605)
Germany, Augsburg, 1583
Parcel gilt silver; enameled plaques;
engraving
H: (ewer) 25 cm (9⅞ in.); diam. (basin)
50.5 cm (19⅞ in.)
85.DG.33.1–.2

In addition to their practical function—serving perfumed water for
guests to wash their hands during a
meal—this ewer and basin commemorate the marriage of Maria Fugger
to Nikolaus Palffy von Erdöd. The coats of arms of the two families surmounted by the date and a pair of clasped hands appear in enameled medallions
on the top of the ewer, in the center of the basin, and engraved underneath it.
The Fuggers, the most illustrious of German banking and mercantile families,
wanted to form an alliance with the Palffys, since one of the main sources of the
Fugger fortunes was their copper mines in the Palffys' native Hungary.

Abraham I Pfleger is recorded as one of the most important goldsmiths of
sixteenth-century Augsburg. Few of his works have survived. This ewer and
basin reveal a style of unusual purity, restraint, and formal severity.

DISPLAY CABINET

Flanders, Antwerp (?), early seventeenth century

Walnut and oak veneered with ebony, tortoiseshell, coconut (?), and ebonized wood

210 x 158 x 74.5 cm (82¾ x 62¼ x 29⅜ in.)

88.DA.10

Antwerp was the most important center for furniture decoration in the Netherlands around the turn of the seventeenth century. As evidence, this cabinet inventively combines architectural forms with fine sculpted figures. The two front doors are decorated with the allegorical figures of Faith and Hope holding their respective attributes of cross and anchor. Charity, who appears on the front drawer, completes the triad of theological virtues. Behind the four fully sculpted caryatid figures that support the cornice at the very top, a receding cupboard opens to reveal an octagonal mirror surrounded by intricately inlaid geometric patterns. The five herms placed around this cupboard may represent the five senses; they are shown drinking, playing music, and so on. The reference to the five senses could reinforce the worldly function of this piece: the display and delectation of precious objects.

ADRIAEN DE VRIES
Dutch, 1545–1626
Rearing Horse, 1613–1622
Bronze
49 x 55 cm (19¼ x 21⅝ in.)
86.SB.488

Throughout his professional life, de Vries seems to have satisfied the dictates of princely taste, for he was courted as a sculptor by some of the most powerful, discerning patrons of Europe, including Charles Emmanuel I, Duke of Savoy, and Emperor Rudolf II of Prague (r. 1576–1612). De Vries worked primarily in bronze, a medium in which he achieved extraordinary virtuosity, and his bronzes are unusual for their consistently high quality and technical sophistication. The *Rearing Horse*—with its smooth musculature, expertly modeled anatomical details, richly colored, reflective surface, and daring balance of the animal's mass on two points—would have merited a prominent place in a royal or aristocratic collector's cabinet.

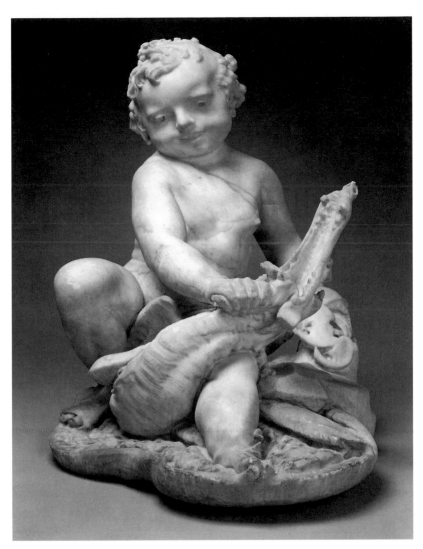

GIAN LORENZO BERNINI
Italian, 1598–1680
Boy with a Dragon, ca. 1614
Marble
54.6 x 43 x 53.3 cm (21½ x 17 x 21 in.)
87.SA.42

Bernini, the greatest, most precocious Baroque sculptor, executed this marble
when he was only about sixteen years old. It is described in early inventories as
representing the young Hercules with a dragon. Unlike its antique precedents,
Bernini has presented the viewer with an unidealized, life-size urchin who
cracks the dragon's jaw not in a heroic struggle but with a very human, mis-
chievous smile. By obscuring the boundary between art and reality, Bernini
invited the viewer's psychological interaction with the sculpture. *Boy with a
Dragon*, fitted for use as an indoor fountain, most likely was commissioned
by Maffeo Barberini (later Pope Urban VIII) to decorate his Roman palace.

BASIN
Italian, Genoa, ca. 1620–1625
Silver
Diam. 75.5 cm (29¾ in.)
85.DG.81

The design of this spectacular basin, executed in exceptionally high relief, is closely based on an oil sketch by the Genoese painter Bernardo Strozzi, now in the Ashmolean Museum, Oxford. It may have been translated into precious metal by a Dutch or Flemish silversmith working in Genoa. As in Strozzi's image, the basin depicts episodes from the story of Anthony and Cleopatra. The organization of these narrative scenes—which include dynamic portrayals of battle, suicide, and death—into concentric bands surrounding a central roundel confirms the basin's Genoese origin. Other silver basins—whose attribution to Genoese workshops is firmly established by that city's *turretta* (turret), with which they are stamped—exhibit a similar disposition of figural reliefs. As in the Museum's example, these primary areas are separated by elegant bands of repeating decorative motifs.

By the seventeenth century Genoa had become one of the most important Italian centers for the production of precious metalwork, and a great quantity of early Genoese silver must once have existed. Despite this, very few secular examples survive from this period; scarcely more than a dozen of high quality are known. Among these the Museum's basin is the largest and most technically daring in its remarkable depth of relief and precision of chasing.

DISPLAY CABINET

Germany, Augsburg, ca. 1620–1630
Ebony, chestnut, walnut, pearwood,
and boxwood; ivory, marble, and
semiprecious stones; enamel;
snakeskin and tortoiseshell
73 x 58 x 59 cm
(28 3/4 x 22 13/16 x 23 1/4 in.)
89.DA.28

This architecturally inspired collector's cabinet was likely influenced by
the projects of Ulrich Baumgartner,
the most prominent Augsburg cabinetmaker of the early seventeenth
century. It displays restrained proportions; elegantly mannered carving in
its so-called Auricular-style, or earlobelike, details above the front "doors"; and
a sensitive colorism contrasting ebony with semiprecious stone inlay. All four
sides open to reveal a surprisingly complex series of drawers and compartments. The biblical, mythological, and historical subjects represented on the
inside were executed in a variety of techniques and materials.

FERDINANDO TACCA

Italian, 1619–1686
Putto Holding Shield, 1650–1655
Bronze
H: 65 cm (25 5/8 in.)
85.SB.70.1

This bronze putto or angel is one of a
pair commissioned in 1650 to adorn
the high altar of the church of San
Stefano al Ponte in Florence. It tes-
tifies to Ferdinando's importance in
the transition of the Florentine sculptural style from the late Mannerism
of Giambologna (see pp. 206–207) to
the Baroque. The putto's elegant gesture and the sway of his hip are the last
remnants of Giambologna's legacy,
which Ferdinando inherited along
with that master's workshop in the
Borgo Pinti. However, the infant's
realistic anatomy, the dramatic play of
light across the complex surfaces of
his hair and drapery, and the resulting
sense of animation underscore Tacca's
mastery of the Baroque idiom.

SIDE TABLE

Italian, Rome, ca. 1720

Gessoed and gilded limewood; modern marble top

93.9 x 190.5 x 96.5 cm (47 x 75 x 38 in.)

82.DA.8

This table is carved with masks, deep scrolls, and female heads in keeping with the exuberant, curvaceous forms of the early eighteenth-century Baroque. It is one of a pair of tables, the other of which is in the Palazzo Barberini, Rome. Although it is not known whether the Barberini table is in its original setting, both pieces undoubtedly were made for one of the grander eighteenth-century palaces. They probably would have supported works of art—such as bronze sculptures or Chinese porcelains—for display.

FRANÇOIS GIRARDON

French, 1628–1715

Pluto Abducting Proserpine, cast ca. 1693–1710

Bronze

H: 105 cm (41⅓ in.)

88.SB.73

Girardon's signed abduction group forms a pair with Gaspard Marsy's *Boreas Abducting Orithyia*, also in the Museum's collection (88.SB.74). Both bronzes are based on models for monumental marble groups commissioned by King Louis XIV to decorate the Parterre d'Eau in the gardens at Versailles. In executing their models, the sculptors faced the artistic challenge of achieving a complex composition viewable from all angles and depicting violent action, all within the boundaries of decorum dictated by the French classicizing style. Their success must have been apparent to contemporary collectors, for bronze reductions of the famous Parterre groups already were in widespread demand by the end of the seventeenth century. Girardon's *Pluto Abducting Proserpine* is the finest and most beautifully patinated of the surviving large casts of his original model.

WALL PLAQUE

Made by Francesco Natale Juvara
(1673–1759)
Italian, 1730–1740
Silver, gilt bronze, and lapis lazuli;
wood backing
70 x 52 cm (27 7/16 x 20 1/2 in.)
85.SE.127

Juvara, son of the Messinese goldsmith Pietro and brother of the
architect Filippo, established a reputation as a maker of fine liturgical
metalwork. His wall plaques, altar
frontals, monstrances, and chalices
once graced the interiors of many
Roman and Sicilian churches. This
plaque represents the Virgin of the
Immaculate Conception as she tram-
ples a snake, symbol of sin.

SIDE TABLE

Italian, ca. 1760–1770
Carved and gilt wood; marble top
105 x 153 x 74 cm (41 5/16 x 60 1/4 x 29 1/8 in.)
87.DA.135

The maker of this rare six-legged table was influenced by the furniture designs
of Giovanni Battista Piranesi, one of the principal forces behind the birth and
development of the Neoclassical style in Europe. As in Piranesi's designs, this
table combines antique ornaments with flamboyant, complex, and curvilinear
elements. Against a wall in an eighteenth-century palazzo, this table would have
served to display decorative objects or sculpture.

ONE OF A PAIR OF VASES

Made by the factory of Geminiano
Cozzi (active 1764–1812)
Italian, Venice, 1769
Hybrid soft-paste porcelain
30.5 x diam. (max.) 24.8 cm
(12 x 9¾ in.)
88.DE.1.1–.2

Produced in one of the leading Italian
porcelain factories of the eighteenth
century, these vases are remarkable for
their large size, unusual shape, elaborate markings, and delicately painted,
sophisticated pictorial scheme. The
painted decoration on the pair, which
copies contemporary print sources,
celebrates the beauty and sovereignty
of the Venetian republic. One vase
displays the sea god Neptune on one
side and an invented riverside town on
the other, while the second vase shows
the allegorical figure of Venice and,
on the other side, a panorama of the
Piazzetta San Marco.

CLODION (Claude Michel)

French, 1738–1814
*Vestal Presenting a Young Woman at the
Altar of Pan*, ca. 1770–1775
Terracotta
H: 45 cm (17¾ in.)
85.SC.166

Through his technical brilliance
and virtuoso handling of wet clay,
Clodion raised to new heights the
aesthetic quality of the terracotta as
an independent work of art rather
than a preparatory sketch or model
for a sculpture in a more permanent
material. The Museum's terracotta
depicts the enactment of a ritual before
a herm of Pan, the god of the fields
associated with lust. The object's
playful, romantic attitude toward the
classical world is typical of the work
of Clodion, who preferred genre
scenes involving marginal mythological figures to the epic dramas of
the principal Olympian gods.

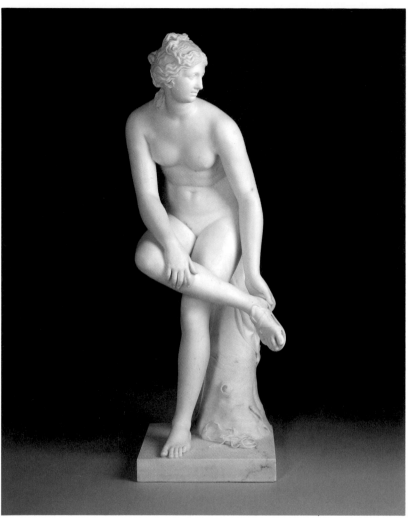

JOSEPH NOLLEKENS
English, 1737–1823
Venus, 1773
Marble
127 x 50.8 x 50.8 cm (50 x 20 x 20 in.)
87.SA.106

Nollekens studied in Rome from 1762 to 1770, and his style, a mannered classicism inflected by coy charm, exhibits the influence of ancient and sixteenth-century Italian sculpture. *Venus*, one of a group of female deities by Nollekens in the Museum's collection, was executed soon after the artist's return to London. With the other figures of Juno and Minerva, the *Venus* formed part of a series sculpted for Lord Rockingham to accompany a marble *Paris*, which Rockingham already owned and believed to be antique. The four statues together illustrate the story of the shepherd king who was empowered to judge which goddess was the fairest. Nollekens chose to depict each of the goddesses in a different state of undress. Venus, the winner, is nude except for the single sandal she is removing.

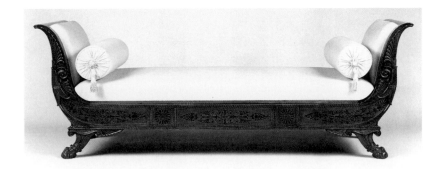

DAYBED

Made by Filippo Pelagio Palagi (1775–1860); inlaid decoration by Gabrielle
Capello (active early–mid-nineteenth century)
Italian, Turin, designed 1832–1835
Maple inlaid with mahogany
80 x 224 x 69 cm (31½ x 88⅛ x 27⅛ in.)
86.DA.511

This lounging bed was executed for the royal bedroom in the Palazzo Racconigi
outside Turin for Carlo Alberto, King of Sardinia (later King of Italy; r. 1831–
1849). Its form and decoration draw on ancient Roman and Napoleonic proto-
types and probably had imperial associations for the designer and his patron.
Architect, portrait painter, ornamentalist, collector, and furniture designer,
Palagi developed an interest in archaeology from his early years in Rome. He
inventively and eclectically combined Egyptian, Greek, Etruscan, and Roman
motifs in his furniture and ornamental designs. The sophisticated inlaid decoration was carried out by Palagi's assistant, an innovator in inlay technique.

JEAN-BAPTISTE CARPEAUX

French, 1827–1875
Bust of Jean-Léon Gérôme, 1872–1873
Marble
H: 60 cm (23⅝ in.)
88.SA.8

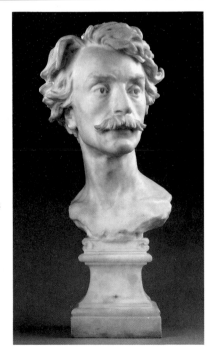

Portrait sculptor to Napoleon III
(r. 1852–1871) and a key figure in the
history of nineteenth-century
sculpture, Carpeaux first modeled a
bust of his friend Gérôme, a prominent painter, in 1871, when the two
were exiles in London during the Paris
Commune. Although numerous versions survive in bronze and plaster,
the Museum's bust is the only known
marble example of this portrait. In it
Carpeaux achieved a romantic image
of the alienated, creative spirit in tur-
moil by accentuating the sunken eyes
and cheeks and giving full play to the
unruly mass of hair.

PHOTOGRAPHS

In mid-1984 the Museum established a new curatorial department dedicated to the art of photography. The opportunity to acquire several of the most important private collections of photographs in the world may be compared to the establishment of the Department of Manuscripts through the acquisition of the finest gathering of illuminated manuscripts then in private hands. The Museum decided to form a photography collection for reasons similar to those advanced in favor of collecting manuscripts and drawings: photography is an art fundamental to its time in which individual works of great rarity, beauty, and historical importance have been made.

Among the collections acquired in their entireties were those of Samuel Wagstaff, Arnold Crane, Bruno Bischofberger, and Volker Kahmen/George Heusch. The gathering of these collections, along with other block acquisitions made at the same time and a continuing program of individual acquisitions, has brought to Los Angeles the most comprehensive corpus of photographs on the West Coast.

The photographs reproduced in this *Handbook* represent a survey of some of the strengths of the collection, which is particularly rich in examples dating from the early 1840's and which includes major holdings by William Henry Fox Talbot, David Octavius Hill and Robert Adamson, Hippolyte Bayard, and other early practitioners who worked around Talbot in England and in France. The collection also includes significant works by some of the most important photographers of the first half of the twentieth century. International in scope, its guiding principle is the belief in the supremacy of certain individual master photographers and in the timeless importance of individual master photographs.

For conservation reasons, photographs, like manuscripts and drawings, cannot be kept on permanent display. At the present time the collection is available to the public by appointment in a study room located in the J. Paul Getty Center for the History of Art and the Humanities, Santa Monica. Rotating exhibitions from the collection are shown in the photographs gallery at the Museum.

CHARLES R. MEADE
American, 1827–1858
Portrait of Louis-Jacques-Mandé Daguerre, 1848
Daguerreotype
16 x 12 cm (6⁵/₁₆ x 4¹³/₁₆ in.)
84.XT.953.1

UNIDENTIFIED PHOTOGRAPHER
Edgar Allan Poe, late October 1848 (?)
Daguerreotype
12.2 x 8.9 cm (4¹³/₁₆ x 3¹/₂ in.)
84.XT.957

For reasons that are not entirely clear, relatively few daguerreotypes of famous persons have survived from the 1840's. Daguerre himself, for example, the father of his profession, was photographed a mere handful of times. The most important of his portraits to survive were made by a stranger, the American Charles Meade. Meade was one of the most prominent daguerreian portraitists

in New York and traveled to France specifically to gain an audience with Daguerre. The Getty portrait is one of five by Meade that are known and is typical in its lighting and in the serene pose of the sitter, whose elbow rests on the table and whose hands are clasped one over the other.

The art of the daguerreotype is one in which the identity of the sitter has often come to have more importance than the name of the maker. This is particularly true in the case of Edgar Allan Poe. Four of the six times that Poe is documented to have posed for a daguerreotypist occurred within the last year of the writer's life. The Getty *Poe* was presented by him to Annie Richmond, one of two women to whom he made romantic declarations in the months following the death of his wife in early 1847. Poe, who has been described as a libertine, a drug addict, and an alcoholic, is represented here as an individual who might well have been the victim of his own unharnessed emotions.

WILLIAM HENRY FOX TALBOT
British, 1800–1877
Oak Tree, mid-1840's
Salt print from paper negative
22.5 x 18.9 cm (8⅞ x 7⁷⁄₁₆ in.)
84.XM.893.1

Talbot's most important invention was one that is easily taken for granted today: the negative from which faithful replicas can be produced. He patented this procedure under the trade name "calotype." Talbot intended his invention to be clearly distinguished from the daguerreotype. Daguerre's procedure resulted in pictures on metal plates that could not be multiplied easily. Daguerreotypes were used almost exclusively for studio portraiture since sitters generally required but a single example, while calotypes required less fussy procedures and therefore were favored when a particular subject had an audience of more than one. Early photographers favored Talbot's process when they traveled for landscape work.

There is no known surviving early daguerreotype of a single tree. Yet trees were a favorite subject for photographers who were influenced by the aesthetic of picturesque romanticism evident in Talbot's treatment of this image.

HIPPOLYTE BAYARD
French, 1801–1887
Arrangement of Specimens, ca. 1841
(from Bayard Album)
Direct positive print
27.7 x 21.6 cm (10¹⁵⁄₁₆ x 8½ in.)
84.XO.968

Bayard was a friend of Daguerre who recognized that the daguerreotype process was flawed: the image could not be multiplied on paper with satisfactory fidelity to the appearance of the original. Inspired by Talbot, Bayard independently devised a way to make paper photographs; however, like the daguerreotype, they were one-of-a-kind objects and could not be replicated faithfully. Bayard's elegant arrangement of plant specimens, textile fragments, and a feather is an experimental forerunner of a working method that the twentieth-century photographer Man Ray thought he invented.

ANNA ATKINS
British, 1799–1871
Pink Lady's Slipper, Collected in Portland (Cipripedium), 1854
Cyanotype
25.8 x 20.2 cm (10³/₁₆ x 7¹⁵/₁₆ in.)
84.XP.463.3

Among the women who played an important role in the development of photography, Atkins was the first to create an extensive body of work. She came to
the new medium through her father, John George Children, who presided over
a February 1840 meeting of the Royal Society at which William Henry Fox
Talbot (see p. 224) disclosed the workings of the positive-negative process.
Children was friendly with Sir John Herschel, who instructed him and Atkins
in the production of cyanotypes. Atkins combined Herschel's and Talbot's
methods to create a visual lexicon of British ferns, algae, and plants, arranging
her specimens in contact with transparent, light-sensitive paper and exposing
the sheets to achieve cameraless negatives. From these negatives editions of
positive cyanotype prints were made.

DAVID OCTAVIUS HILL AND ROBERT ADAMSON
Scottish, 1802–1870; 1821–1848
Newhaven Fisherman, 1844–1848
Calotype
20.7 x 15 cm (8^{1}/$_{16}$ x 5^{15}/$_{16}$ in.)
84.XM.445.1

Hill brought his abilities as a portrait painter to the partnership in photography
he formed with Adamson in 1844. Adamson brought his skills in the manipula-
tion of the calotype process, which he had learned indirectly from its inventor,
William Henry Fox Talbot. In the four years before Adamson's death the
partners produced the first corpus of photographs made as art rather than
experiment. Their prints are characterized by a rembrandtesque chiaroscuro in
tonalities created by gold chloride, which also gave the prints permanence.

NADAR (Gaspard-Félix Tournachon)
French, 1820–1910
Self-Portrait, ca. 1855
Salt print
20.5 x 16.9 cm ($8^{11}/_{16}$ x $6^{5}/_{8}$ in.)
84.XM.436.2

Nadar turned his back on the tradition of portraiture established by the daguerreian generation. Daguerreotype portraits were generally formal and hieratic,
in contrast to the informality and naturalism that Nadar wished to express and
that are typified by this self-portrait. The word *candid,* though not in use in connection with photography in the 1850's, best describes Nadar's point of view.
He made numerous self-portraits in order to experiment with various poses
and gestures.

   Nadar always posed his subjects against the same neutral background and
placed them below a skylight that cast gentle light from above and left delicate shadows under the eyes, around the nose, and in facial folds and hollows.
His ultimate goal was to create a corpus of photographs representing the most
celebrated citizens of Paris in the arts, literature, and the professions during the
Second Empire. He called this undertaking the *Panthéon Nadar.*

JULIA MARGARET CAMERON
British, 1815–1879
Ellen Terry, 1864; carbon print, 1872
Diam. 24.2 cm (9½ in.)
86.XM.636.1

Julia Margaret Cameron is renowned for her portraits of famous men such as
Thomas Carlyle, Alfred, Lord Tennyson, and Sir John Herschel as well as of
women friends and relatives. This image of Ellen Terry, however, is one of her
few known photographs of a female celebrity. Terry, the popular child actress of
the British stage who later became Dame Ellen Terry, was only sixteen years old
when Cameron photographed her. She had just married the eccentric painter
George Frederic Watts, who was thirty years her senior and an artistic mentor
to Cameron. The ill-fated union lasted less than a year.

Cameron's portrait echoes Watts' study of Terry entitled *Choosing* (1864;
England, private collection). Here, as in the painting, Terry is shown in profile
with her eyes closed, an ultrafeminine, ethereal beauty in a melancholic dream
state. In this guise she embodies the Pre-Raphaelite ideal of womanhood rather
than the wild, boisterous teenager she was known to be. Terry's enchanting
good looks, if not her personality, suited this ideal perfectly. The round format
of the photograph, referred to as a tondo in painting, was very popular among
Pre-Raphaelite artists.

Cameron commissioned the London Autotype Company to make carbon
prints of this image. These prints are more permanent than photographs
because they are composed of pigment rather than light-sensitive silver salts.

ALEXANDER GARDNER
American, 1821–1882
Lincoln on the Battlefield of Antietam, Maryland, 1862
Albumen print
22 x 19.6 cm (8⅝ x 7¾ in.)
84.XM.482.1

Lincoln was the first American president to recognize the importance of photography and to make time in his busy life to be photographed on many occasions. Most of the surviving photographs of him were made in Washington, D.C., or in village studios in Illinois. Here we see the commander-in-chief conferring with Major Allan Pinkerton, chief of the Secret Service, and Major General John McClernand in the combat zone.

The genius of this photograph lies in Gardner's ability to build a composition around the intrusive details of camp life. The tent and tent lines dominate the composition. Thus the viewer's eye is drawn as much to the fastenings on the lines as it is to the faces of the principals. Despite compositional interruptions, however, the statuesque figure of Lincoln remains the center of interest.

GUSTAVE LE GRAY
French, 1820–1882
Cavalry Maneuvers, Camp at Châlons, 1857
Albumen print
31 x 36.7 cm (12³/₁₆ x 14⁷/₁₆ in.)
84.XO.377.12

In 1857 Napoleon III commissioned Le Gray, then at the height of his photographic career, to commemorate the inauguration of, and chronicle life in, the vast military camp established that year on the plain at Châlons-sur-Marne. Designed to accommodate twenty-five thousand imperial guards and staff, the camp spread over thirty thousand acres of flat terrain. Le Gray's job was to provide grandeur to the site and the activities taking place there. His photographs comprise images of cavalry exercises on a grand scale, genre studies of troops in bivouac, formal portraits of officers, records of ceremonies including High Mass, and an overall panorama of the camp as seen from the emperor's central pavilion. In this photograph, lines of cavalry behind a field bulwark are cloaked in an atmospheric mist new to photography. They occupy a narrow band across the center of the image, leaving a great empty swath of pale sky and a dark, nearly vacant foreground. Although many of these soldiers recently had returned from real war in the Crimea, Le Gray romanticized their maneuvers as if they were a corps de ballet behind a scrim. (Given the sweeping nature of cavalry tactics at the time, a certain distance between camera and subject may have been a practical necessity, of course.) Napoleon III gathered Le Gray's photographs into elaborate albums for presentation to high-ranking staff officers such as Commander Verly, who was the first owner of the album in the Museum's collection in which this photograph is found.

ROGER FENTON
British, 1819–1869
Costume Study, 1858
Albumen print
45 x 36.3 cm (17 $^{11}/_{16}$ x 14 $^{1}/_{4}$ in.)
84.XP.219.32

Even though his career as a photographer lasted only ten years, Roger Fenton had a strong impact on the history of the medium. A founder of the Photographic Society of London (now the Royal Photographic Society), Fenton is best known for his images of the Crimean War, among the earliest applications of photography to war reportage. He also made photographs that represent fictional situations. He often drew inspiration for subject matter and composition from painting, which he had studied before turning to photography. In creating this image Fenton looked to the Orientalist works of, among others, Eugène Delacroix. These European depictions of the colonized Islamic world were romanticized inventions rather than accurate representations of Middle Eastern life. Fenton's harem scene is pure fiction: an English studio provided locale, two Westerners modeled as musicians, and the dancer's graceful pose was achieved with the aid of wires.

CARLETON E. WATKINS
American, 1829–1916
Columbia River, Oregon, 1867
Albumen print
40.5 x 52.3 cm (16¹/₁₆ x 20¹¹/₁₆ in.)
85.XM.11.2

While the Civil War was raging in the Eastern United States, Westerners, still buoyant from post–Gold Rush prosperity, continued to lead normal lives. The best Eastern contemporaries of Watkins—Gardner, O'Sullivan, and Russell— spent their time photographing aspects of the war (see p. 230), while Watkins, who ranks among the greatest American photographers of all time, had the leisure to ripen his style to full maturity between 1861 and 1868. Watkins had the gift to create complex compositions from very simple motifs and the power of perception to apprehend ephemeral forces in nature that form a seamless web of formal relationships. The three key elements of this picture are the massiveness of the rock formations at either side, the transient quality of the boat loaded with a box of enormous apples, and the delicacy of the light reflected from the water's surface.

Watkins designed photographs brilliantly to achieve a painterly interplay between surface pattern and spatial dimensions. The network of intricately connected compositional elements evident in *Columbia River, Oregon* is chiefly responsible for the picture's palpable sense of space and is typical of this concern. Watkins' photographs were used as reference sources by painters such as Thomas Hill and Albert Bierstadt.

Watkins was also an excellent technician who worked in a variety of materials. He frequently made stereographs, a type of miniature photograph that functioned for him as a sketching medium. After visualizing his subject, he would proceed to make mammoth plate negatives that yielded the presentation prints for which he was most celebrated.

THOMAS EAKINS
American, 1844–1916
Eakins' Students at the Site for "The Swimming Hole," ca. 1883
Albumen print
16.5 x 12.2 cm (6½ x 4¾ in.)
84.XM.811.1

Photography was invented just when painters seemed to require a new way of seeing the world. Although many nineteenth-century painters dabbled in photography, very few carried their experiments far enough to produce a significant corpus of work. Two painters who did, and who are still universally respected, were Thomas Eakins and Edgar Degas (see p. 235).

Eakins began to photograph in the late 1870's. Interested in the poses and gestures of static nude figures, he made this study of seven young men at a swimming hole around 1883. Male nudes were then very uncommon in photography, and Eakins was among the first to experiment with this subject. His models were students at the Pennsylvania Academy of the Fine Arts, which censored him for this practice. Eakins used this photograph as a model for a painting now in the Amon Carter Museum, Fort Worth.

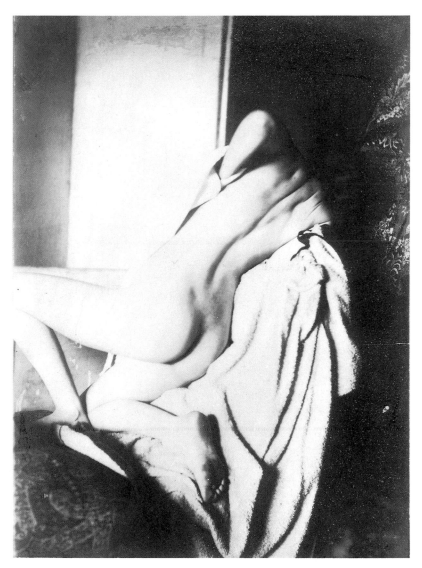

EDGAR DEGAS
French, 1834–1917
After the Bath, Woman Drying Her Back, 1896
Gelatin silver print
16.5 x 12 cm (6^{11}/$_{16}$ x 4^{3}/$_{4}$ in.)
84.XM.495.2

Degas began to photograph in the early 1880's, toward the end of a career distinguished by persistent experimentation in etching, lithography, monotype, and, finally, photography. He used the latter as a medium of creative expression and as a tool in the preparation of drawings and paintings. This model's highly contorted pose, unlike the relaxed posture of Degas' related studies of other nudes, is the prototype for a painting (Philadelphia Museum of Art).

CHARLES SHEELER
American, 1883–1965
Doylestown House—The Stove, 1917
Gelatin silver print
22.9 x 16.2 cm (9 x 6⅜ in.)
88.XM.22.1

In 1910 Charles Sheeler moved into a small farmhouse in Bucks County,
Pennsylvania, that became the subject for a suite of photographic studies.
Sheeler's spare and unconventional photograph *Doylestown House—The Stove*
unites rustic Americana and hard-edged modernism. By carefully illuminating
the composition with light flooding from the stove, Sheeler reconciled the
contrast between the rough textures of the materials comprising the farmhouse
and the crisp, metallic silhouette of the stove itself.

PAUL STRAND
American, 1890–1976
Portrait—New York, 1916
Platinum print
34.3 x 25.1 cm (13½ x 9⅞ in.)
89.XM.1.1

About the time he was creating abstractions from architectural elements and arrangements of cups and bowls, Strand also was photographing people in the streets of New York. He attached a false lens to his reflex camera to photograph his anonymous subjects unawares and achieve an unself-conscious naturalism. Strand enlarged his negatives and printed them in rich brown tones of platinum. Alfred Stieglitz devoted the last issue of his journal *Camera Work* to eleven Strand photographs, including this one, the only surviving platinum print from the negative.

AUGUST SANDER
German, 1876–1964
Frau Peter Abelen, Cologne, 1926
Gelatin silver print
23 x 16.3 cm (9¹/₁₆ x 6⁷/₁₆ in.)
84.XM.498.9

Sander spent the greater part of his career creating an ambitious series of
photographs entitled *Menschen des 20. Jahrhunderts.* His goal was to describe
as truthfully as possible every aspect of German society. This portrait of the
wife of the painter Peter Abelen is full of ambiguity. If the title were unknown,
one would wonder whether a male or female was represented and whether
the face expressed hate or longing.

DORIS ULMANN
American, 1882–1934
Portrait Study, South Carolina, ca. 1929–1930
Platinum print
20.6 x 15.4 cm (8⅛ x 6¹/₁₆ in.)
87.XM.89.81

Ulmann, a small, frail woman from Manhattan, began her lifelong project of photographing American "types" with portraits of the Dunkards, Shakers, and Mennonites in rural Pennsylvania and other Eastern states. Through her circle of literary friends in New York, she met the South Carolina writer Julia Peterkin, who persuaded her to produce illustrations for *Roll, Jordan, Roll* (1933), Peterkin's book documenting the customs and inhabitants of Lang Syne Plantation. This portrait of an adolescent girl probably was made as part of that project. It displays both the influence of the Pictorialist Clarence White and Ulmann's technique of photographing in natural light with a view camera and soft-focus lens.

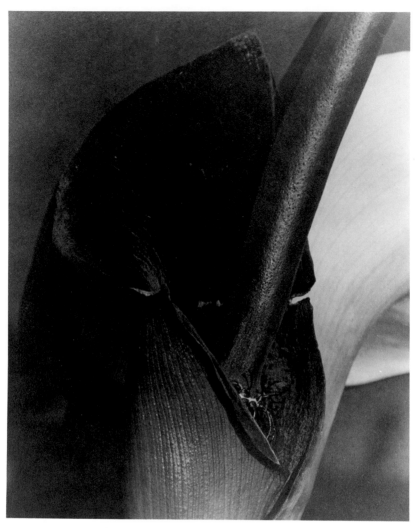

IMOGEN CUNNINGHAM
American, 1883–1976
Black and White Lily, 1928/29
Gelatin silver print
30 x 23.4 cm (11⅝ x 9³⁄₁₆ in.)
84.XP.208.1

We generally have little difficulty in determining when a painting or sculpture
is the product of its maker's imagination. Imagination is a more difficult con-
cept to grasp in the art of photography and generally involves seeing familiar
objects in unexpected ways. In this photograph, first shown in the 1929 exhibition *Film und Foto* in Stuttgart (catalogue number 165), Cunningham used
imagination to retain a delicate balance between recognizable actuality and styl-
ization. The print challenges our ability to decipher its subject by eliminating
the white range of the tonal scale, thus blurring the edges between shapes in a
way not unlike a painting made with a very broad brush.

EDWARD WESTON
American (1886–1958)
Two Shells, May 1927
Gelatin silver print
24 x 18.4 cm (9½ x 7¼ in.)
88.XM.56

Two Shells is one of the finest in a series of close-up still lifes made by Weston in 1927. In this seminal group, which led to his famous Pepper series of 1929, Weston tried to portray "the very quintessence of the thing itself," a phrase he used repeatedly in his *Daybooks* to connote not only the physical essence of a thing but also its inner spirit, or "life force."

This stark, sensual close-up of two nested nautilus shells may be compared to paintings from the same period by Georgia O'Keeffe, whose work Weston admired in New York in 1922. To create this dramatic, sensitive composition, in which ordinary shells become radiant sculpture, Weston used an eight-by-ten-inch camera and an exposure time of at least three hours.

ALBERT RENGER-PATZSCH
German, 1897–1966
Flatirons for Shoe Manufacture, ca. 1928
Gelatin silver print
23 x 17 cm (9¹/₁₆ x 6¹¹/₁₆ in.)
84.XM.138.1

Born in Würzburg, Renger-Patzsch concentrated on classical studies at the
Kreuzschule, Dresden, served in the German army, and studied chemistry
before setting up his own commercial photography business. His 1928 land-
mark publication, *Die Welt ist schön,* reveals a distinctive, voracious view of the
variety to be found in manmade and natural structures. It also reflects Renger's
sympathy for the New Objectivity movement, founded by German painters of
the 1920's. *Flatirons* illustrates Renger's success in emphasizing the materiality of
commonplace objects.

ALEXANDER RODCHENKO
Russian, 1891–1956
En Route, 1932
Gelatin silver print
18 x 12 cm (7 ⅛ x 4 ¾ in.)
84.XM.258.1

During the decade following the Russian Revolution of 1917, avant-garde
experiments by important artists were briefly thought to reflect the ideals of
the newly created state. Photographers experimented with both worm's-eye
and bird's-eye points of view, as well as with telephoto lenses, extreme enlarge-
ments, and action-filled subjects such as the political rally represented here.
This photograph functions as both art and political propaganda.

ANDRE KERTESZ
American (b. Hungary), 1894–1985
Chez Mondrian, 1926
Gelatin silver print
10.9 x 7.9 cm (4⁵⁄₁₆ x 3⅛ in.)
86.XM.706.10

The Hungarian-born Kertész once described himself as a "Naturalist-
Surrealist." In his most characteristic photographs, an abiding interest in the
prosaic aspects of life blends with a surrealistic perspective. His work is marked
by an ability to surprise us by addressing familiar subjects in unusual ways. If
the subject was a still life, Kertész chose his viewpoint deftly and occasionally
made a subtle alteration to gain his desired effect. He described for a friend the
shaping of this particular composition: "The door to [Mondrian's] staircase
was always shut, but as I opened it in my mind's eye the two sights started to
present themselves as two halves of an interesting image that I thought should
be unified. I left the door open, but to get what I wanted I had to move a sofa."

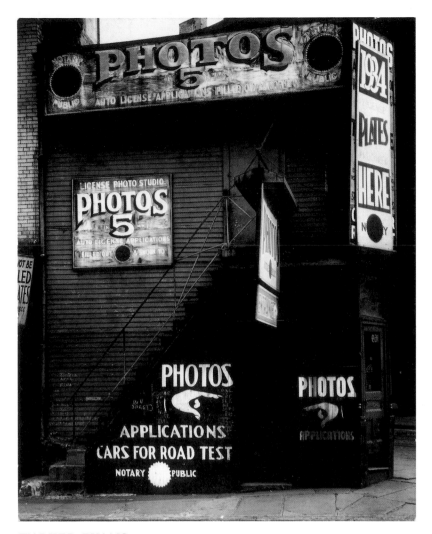

WALKER EVANS
American, 1903–1975
License Photo Studio, New York, 1934
Gelatin silver print
18.2 x 14.4 cm (7³/₁₆ x 5¹¹/₁₆ in.)
84.XM.956.456

Best known for his images of Depression-era rural life, Walker Evans was a pivotal figure in American documentary-style photography. In 1938 his seminal work, *American Photographs*, was published. This book includes eighty-seven photographs of people, architecture, and vernacular art made over a decade in the Eastern United States and Cuba. Ostensibly a proclamation of American culture, the book's underlying theme is photography itself. Using the volume's narrative format, Evans presented photographs in a sequence intended to be "read." This view of a run-down photo studio is *American Photographs*' opening image. With it Evans set the tone for the entire book. The harsh frontal view, unidealized subject, and fascination with commercial signs recur throughout.